KU-633-691

ELECTRON MICROSCOPY AND MICROANALYSIS OF CRYSTALLINE MATERIALS

ELECTRON MICROSCOPY AND MICROANALYSIS OF CRYSTALLINE MATERIALS

Edited by

J. A. BELK
B.Sc., Ph.D., F.I.M., C.Eng.

*Metallurgy Branch, The Royal Military College of Science,
Shrivenham, Swindon, UK*

APPLIED SCIENCE PUBLISHERS LTD
LONDON

APPLIED SCIENCE PUBLISHERS LTD
RIPPLE ROAD, BARKING, ESSEX, ENGLAND

British Library Cataloguing in Publication Data

Electron microscopy and microanalysis of
crystalline materials.
1. Electron microscopy   2. Microchemistry
3. Crystallography—Technique
I. Belk, John Anthony
620.1′1299   QD906.7.E37

ISBN 0-85334-816-2

WITH 11 TABLES AND 89 ILLUSTRATIONS

© APPLIED SCIENCE PUBLISHERS LTD 1979

All rights reserved. No part of this publication may be reproduced, stored in
a retrieval system, or transmitted in any form or by any means, electronic,
mechanical, photocopying, recording, or otherwise, without the prior
written permission of the publishers, Applied Science Publishers Ltd, Ripple
Road, Barking, Essex, England

Printed in Great Britain by Galliard (Printers) Ltd, Great Yarmouth

D
620.1129′9
ELE

Preface

The book *Electron Microscopy and Microanalysis of Metals* of which this volume is, in effect, a revised edition, set out to approach the subjects from an essentially practical point of view, giving the theoretical background necessary to enable to reader to carry out useful experiments and to understand the phenomena he observed. The intervening years have required the book to be completely rewritten in order to include the newer developments, but the aim of the text remains unchanged.

The scanning electron microscope now has its own chapter (Chapter 5) and many other new techniques are introduced and explained in chapters whose headings are very similar to those used before. Inevitably there are some topics that occur in more than one chapter but these are cross-referenced and each mention helps to explain and illustrate the topic rather than repeat information that is given elsewhere in the volume.

The field of regular application of the techniques has also widened since 1968 and this is reflected in the change of title and the inclusion of examples taken from other materials besides metals. The text will be useful for postgraduate and final-year undergraduate students who need an introduction to the techniques so that they can understand and use them to study some of the many problems in materials science.

Studies of the influence of stacking-fault energy on the deformation mechanisms of alloys, the complex interactions between precipitates and dislocations, the composition and crystal structure of small components of a microstructure, and their orientation relationships have led to a great increase in the fundamental understanding of the properties of crystalline materials. It is hoped that this volume will introduce newcomers to the field to the techniques used, help them to appreciate the results obtained and allow them to go on to study the techniques further and make contributions in their own right.

J. A. BELK

Contents

Contributors

J. A. BELK, B.Sc., Ph.D., F.I.M., C.Eng.

Metallurgy Branch, The Royal Military College of Science, Shrivenham, Swindon SN6 8LA, Wilts., UK.

C. W. HAWORTH, M.A., D.Phil., M.Inst.P.

Department of Metallurgy, University of Sheffield, Sheffield S1 3JD, Yorks, UK.

W. JAMES, B.Sc.

Department of Physics, University of Aston in Birmingham, Birmingham B4 7ET, UK.

M. H. LORETTO, M.Met., D.Sc.

Department of Physical Metallurgy and Science of Materials, University of Birmingham, P.O. Box 363, Birmingham B15 2TT, UK.

G. W. LORIMER, B.A.Sc., Ph.D.

University of Manchester, Department of Metallurgy, Grosvenor Street, Manchester M1 7HS, UK.

W. J. M. SALTER, B.Sc., M.Sc., Ph.D., D.C.T. (Batt.)

Iron and Steel Industry Training Board, 14 Commercial Street, Sheffield S1 1QP, Yorks., UK.

R. E. SMALLMAN, B.Sc., Ph.D., D.Sc., C.Eng., F.I.M.

Department of Physical Metallurgy and Science of Materials, University of Birmingham, P.O. Box 363, Birmingham B15 2TT, UK.

N. SWINDELLS, B.Sc., M.Sc., Ph.D.

Department of Metallurgy and Materials Science, University of Liverpool, P.O. Box 147, Liverpool L69 3BX, UK.

C. G. VAN ESSEN, M.A., D. Phil.

Patscentre International, Cambridge Division, Melbourn, Royston SG8 6DP, Herts., UK.

1

Principles of Electron Optics

W. JAMES

University of Aston in Birmingham, Birmingham, UK

1.1 INTRODUCTION

The focussing of a beam of electrons by an electromagnetic field was first demonstrated in 1926. The equations of motion of an electron in a field of axial symmetry are readily deduced from Hamilton's principle and an analogy between Hamilton's principle and Fermat's principle enables us to describe the properties of an electric or magnetic field in terms usually applied to a light optical system, viz. the cardinal points. We therefore speak, by analogy, of 'electron optics' and of an 'electron lens'.

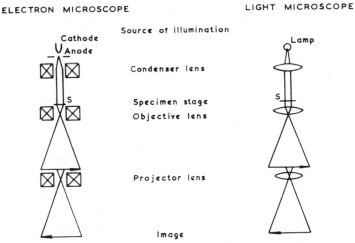

Fig. 1.1.   Comparison of transmission electron microscope and light microscope.

The electron microscope has a source of electrons with condenser lenses, objective lens and projector lens performing the same functions as their light-optical counterparts (Fig. 1.1). However, it should be emphasized that there are some very important differences between light microscopy and electron microscopy. The electron microscope must be evacuated so that the electrons pass through the system unimpeded by residual gases and this imposes certain technical restrictions. But the most important difference lies in the nature of contrast formation, which in light optics is usually by absorption, but which in electron microscopy is by the complex processes of electron scattering or diffraction. The most important of these is nuclear scattering and therefore such contrast effects will not be simply related to the chemical structure of the specimen.

1.2   LIMIT OF RESOLUTION

The limit of resolution of an optical microscope is determined principally by the shortest wavelength of visible light available. This gives a limit of resolution of about 200 nm. The effective wavelength (in nanometres) of an electron beam is given by

$$\lambda = \left\{ \frac{1 \cdot 5}{V(1 + 0 \cdot 978 \times 10^{-6}V)} \right\}^{1/2}$$

where V is the accelerating voltage. Thus, acceleration through a modest potential difference of 150 V produces an effective wavelength of 0·1 nm. However, the electron beam must be sufficiently energetic to penetrate the specimen and this means that voltages of 50–100 kV are usual, giving a wavelength of about 0·004 nm. Other factors such as lens aberrations prevent this short wavelength being fully exploited and the resolving power of most high performance microscopes lies between 0·15 nm and 1 nm.

1.3   GEOMETRICAL OPTICS

Figure 1.2 shows the cardinal points of an optical system. The principal planes, P, P^1 and the focal planes, F, F^1 define the imaging properties of the lens for paraxial rays. When the cardinal points have been found, the image distance is calculated from

$$\frac{1}{l^1} - \frac{1}{l} = \frac{1}{f^1} = -\frac{1}{f}$$

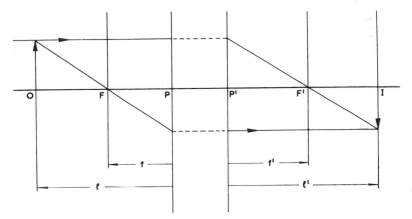

Fig. 1.2. Optical system showing cardinal planes.

and the magnification from

$$\frac{l^1}{l} = 1 - \frac{l^1}{f^1} \eqsim -\frac{l^1}{f^1}$$

When non-paraxial rays are included, an imperfect image is formed, the deficiencies being attributed to the aberrations of the lens. The most important are spherical aberration, astigmatism and chromatic aberration. The light-optical designer can minimize selected aberrations but the electron optical designer is limited by the nature of the electromagnetic fields (he cannot, for example, make a diverging lens) and his only method of reducing the aberrations is (a) by producing a monochromatic radiation by stabilizing the voltage and current supplies, and (b) by reducing the aperture of the lenses. Thus the limit of resolution calculated from the Rayleigh criterion $d = 0.6\lambda/\alpha$ is limited not by the wavelength of the radiation but by the smallness of the aperture (α). This size of the aperture is a compromise between the loss of image quality due to spherical aberration (astigmatism can be reduced to an arbitrarily small value by the use of auxiliary fields) and that due to the diffraction effects at the aperture. There is some dispute as to the best method of combining these two expressions to calculate the optimum size of aperture. All methods give expressions of the form

$$\alpha_{\text{opt}} = K(\lambda/C_s)^{1/4}$$

and the value of the constant K may vary from 0.4 to 1.4 depending on the methods of calculation.

The radius of the disc of confusion due to spherical aberration referred to the object space is given by $d_s = C_s \alpha^3$ where C_s is the spherical aberration coefficient, and α is the angular aperture of the lens. Thus d_s increases very rapidly as the aperture of the lens is increased. The radius of the disc of confusion due to diffraction at the aperture is given by $d = 0.6\lambda/\alpha$ and thus decreases as α increases.

For most commercial microscopes α_{opt} is about 3×10^{-3} radian and with a lens of focal length 3 mm, this gives a physical aperture of diameter 10 μm. This is rather small for general use and difficult to keep free from contamination, so that in practice an aperture of about 25 μm is commonly used. The resolving power is then about 0.5 nm.

1.4   ELECTRON LENSES

In an electrostatic field the force on an electron is given by $\mathbf{F} = -e\mathbf{E}$ where $\mathbf{E}$ is the electric field, and therefore always perpendicular to the equipotential surfaces. The electrodes generally have axial symmetry about the optic axis of the system. The diagram (Fig. 1.3) shows a section through a two-cylinder lens and the equipotentials. As a first approximation, the electron trajectories can be plotted using Snell's Law in the form $\sin i / \sin r = \phi_2/\phi_1$ at each equipotential, but the errors involved in this method are cumulative and more sophisticated techniques are necessary. However, few electrostatic lenses are now used in electron microscopy (except in the electron gun) because of the difficulty of avoiding contamination and

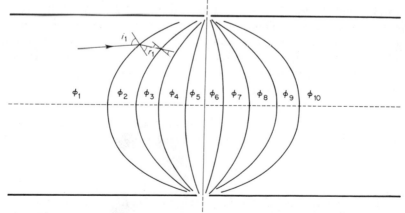

Fig. 1.3.   Electrostatic lens showing electron path traced through.

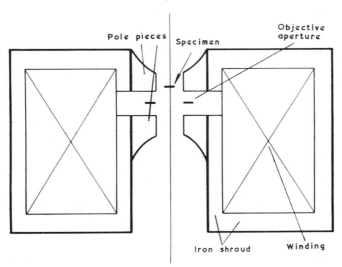

Fig. 1.4. Magnetic objective lens.

electrical breakdown in the relatively poor vacuum conditions in the microscope.

In a magnetic field, the force on an electron is given by $\mathbf{F} = -e\,(\mathbf{v} \times \mathbf{B})$. The force is therefore dependent on the velocity of the electron $\mathbf{v}$ and is perpendicular to the magnetic field $\mathbf{B}$. This produces a rather more complex motion than the electrostatic field but one which can be discussed in terms of the motion in a meridional plane with an effective electrostatic potential derived from the magnetic field. The meridional plane rotates about the axis with an angular velocity depending on the value of the magnetic field at that point. This is known as the *Larmor frequency* and is given by $\omega = eB/2m$. The resulting motion produces a rotation of the image about the axis and this introduces additional aberrations into the system. The magnetic lens consists of a solenoidal coil shrouded in soft iron and with iron pole-pieces to concentrate the field to a small axial volume (Fig. 1.4). The lens is a 'thick' lens and the focal length is not a continuous function of the excitation since the electron paths may cross the axis one or more times, as shown in Fig. 1.5(a). However, lenses are generally operated in the most advantageous manner at a condition corresponding approximately to the first minimum of Fig. 1.5(b). The curves in Fig. 1.5 refer to the usage of the lens as projector lens when the object plane is outside the field. In an objective lens (Fig. 1.6(a)) the specimen is immersed in the field and only that part of the field between the object plane and the image plane is used in image

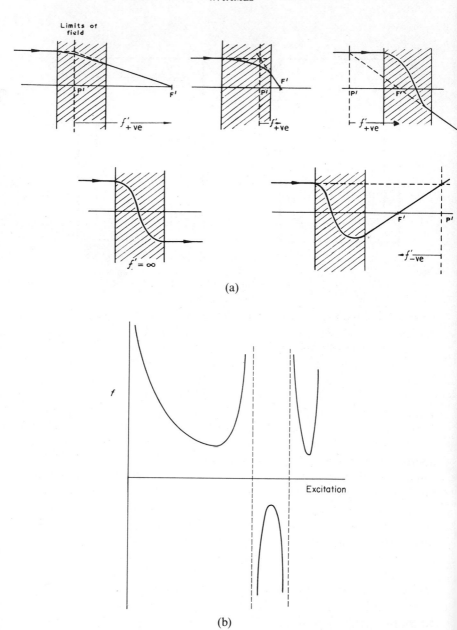

(a)

(b)

Fig. 1.5. (a) Ray paths through projection lens; (b) variations of focal length with excitation.

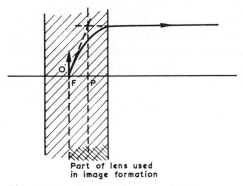

Part of lens used
in image formation

Fig. 1.6(a). Lens used as an objective.

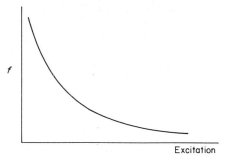

Excitation

Fig. 1.6(b). Variation of focal length with excitation.

formation. Under these conditions the variation of focal length is continuous and decreases with increasing excitation (Fig. 1.6(b)). Lenses of small focal length are obviously very desirable for use as objectives since the aberrations increase with focal length.

The minimum focal length obtainable is determined by geometrical factors and the magnetic saturation of the pole-pieces. The necessary magnification is also obtained without excessive image distance and this results in a smaller and mechanically more stable instrument. A three-stage imaging system also has this desirable effect and is commonly used.

1.5 THE TRANSMISSION ELECTRON MICROSCOPE

1.5.1 The Illuminating System
The electron gun of the electron microscope is analogous to the lamp of the light microscope. It is a three-electrode system of filament, cathode shield

and anode; the latter two components form a converging electrostatic lens which produces an image of the emitting area of the filament (Fig. 1.7). In fact, the cross-over forms a smaller 'effective' source and is used as such. The correct illuminating conditions at the specimen are obtained by the condenser lens system. A single condenser lens of short focal length will produce a de-magnified image of the 'effective' source but if this is to be

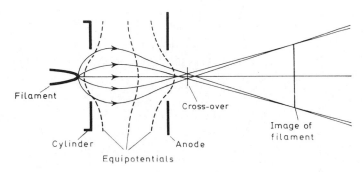

Fig. 1.7.   Electron gun.

projected into the specimen plane, a second condenser lens of longer focal length is required to accommodate specimen air locks, special stages, etc. Contamination of the specimen occurs over the area of irradiation due to the reduction of absorbed layers of carbonaceous vapours, and it is desirable to reduce the contaminated area to a minimum. When the effective source is imaged in the specimen plane, the spot size is a minimum and depends on the aperture of the condenser lens, but the convergence of the beam is maximal. If the second condenser lens is over-focussed the illumination is reduced, and the beam is now limited by the size of the source (Figs. 1.8(a) and (b)). The convergence of the beam is reduced and the phase contrast effects are enhanced.

The electron gun and condenser system together determine the degree of coherence of the electron beam at the specimen. This is of great importance since many of the contrast mechanisms depend on the interference of the diffracted beams. There are two aspects of the coherence condition that have to be considered: the longitudinal or chromatic coherence and the transverse or spatial coherence. The former is determined by the stability of the power supplied and is given by $\lambda E/\Delta E$ where λ is the electron wavelength, E is the energy of the beam and ΔE the fluctuation in the energy. Putting in typical numbers we get a value of about $1\ \mu m$. The

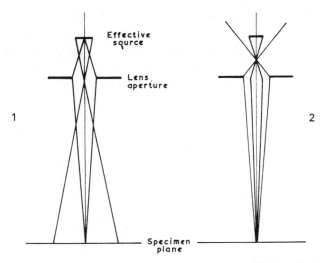

Fig. 1.8(a).   Action of condenser lens. (1) Critical focussing, spot limited by size of aperture. (2) Overfocussed, spot limited by size of source.

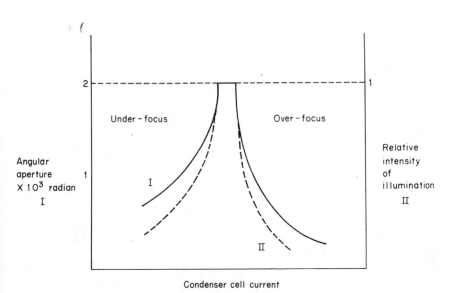

Fig. 1.8(b).   Dependence of angular aperture on excitation.

transverse coherence is determined by the angular aperture of β of the illumination and is approximately given by $\lambda/2\beta$, which yields a typical value of about 2·0 nm. Thus, while the electron microscope specimen can be considered to be illuminated coherently through its thickness, the transverse area so illuminated has a diameter of only about 2·0 nm under normal conditions. (The transverse coherence length can be increased somewhat by using a smaller condenser aperture and overfocussing the second condenser lens.) Thus we can imagine the specimen to be divided into a series of columns of diameter about 2 nm parallel to the optical axis, each of which diffracts independently of its neighbours. This concept is of importance in the theory of contrast (see Chapters 3 and 4).

1.5.2 The Image-forming System

The objective lens is the most important lens of the system since it produces the primary image which is subsequently enlarged by the projector lenses. Great care must therefore be taken in the construction of the pole-pieces and in their maintenance. Contamination or physical damage will lead to excessive astigmatism.

The projector system produces an enlargement of the primary image on the fluorescent screen. Two or more lenses may be used, giving a higher degree of flexibility in covering the range of magnification from about 1500 × up to 200 000 × with minimum distortion. The back focal plane of the objective lens may also be imaged when the Fraunhofer diffraction pattern of the specimen is projected (Fig. 1.9). Adjustable apertures in the intermediate image plane allow diffraction patterns from selected areas of the specimen to be studied. The microscope is completed by a system of viewing and recording the image, i.e. fluorescent screen and camera.

An important consequence of the small angular aperture of the microscope lenses is the large depth of field and the depth of focus. The depth of field is the axial distance in the object space in which all specimen planes are effectively in focus, i.e. the disc of confusion due to displacement from the Gaussian object plane is smaller than the disc of confusion due to other effects such as aberrations and diffraction. The depth of field for an electron microscope is about 2 μm and since the specimen is always less than 500 nm in thickness, the image is a superposition of all the features through the thickness of the object and all are simultaneously in focus. The depth of focus is the axial region in the image space in which the image quality is the same, and is simply related to the depth of field by the square of the magnification. Thus the depth of focus is very large, and thus the positions of the fluorescent screen and the camera are not very critical. When the

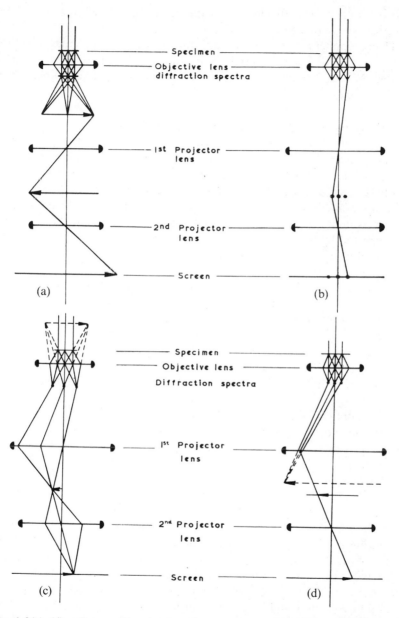

Fig. 1.9(a)–(d). Ray paths in imaging systems. (a) High magnification; (b) diffraction spectra imaged; (c) low magnification; (d) low magnification.

specimen is properly focussed the camera may be placed in a convenient position above or below the screen.

The components of the microscope must be accurately aligned along the optic axis of the system. Logical routines based on the known image movements accompanying changes of current or mechanical displacement enable this lining-up process to be carried out.

1.5.3 Correction of Astigmatism in Magnetic Lenses

The magnetic fields produced by lenses are never perfectly symmetrical about the axis, owing to difficulties of manufacture and inhomogeneities in the iron. The effect is of major importance in the objective lens (where it limits the resolution) and in the illuminating system (where it reduces the intensity of the beam). A variable amount of astigmatism is also produced by contamination of apertures and pole-pieces. Fortunately, the defect can be corrected by the use of small additional fields of variable strength and orientation. The compensation for astigmatism in the objective lens is conveniently carried out by observation of the diffraction effects occurring round the edge of a hole in a carbon film. The specimen consists of a carbon film with a number of small circular holes and is viewed in a slightly over- or under-focussed condition. The image is the Fresnel diffraction pattern around the hole, consisting of alternative light and dark fringes, the thickness of which depends on the position of the plane in focus. The existence of two focal planes separated by the astigmatic distance produces an asymmetry in the circular fringe, which is best seen when the degree of de-focus is at the minimum necessary to produce visible fringes. The compensating fields are then adjusted to produce symmetrical fringes. The separation of the fringes may also be used to estimate the resolving power of the instrument.

1.5.4 Contrast

When a beam of electrons passes through a specimen the electrons undergo elastic and inelastic scattering. The former definition assumes collison but no exchange of energy with the nuclei, and the latter a loss of energy to the bound electrons and/or to the electron plasma of the specimen. In either case, information about the specimen is contained in the scattered (or diffracted) beams. The information contained in the inelastically scattered beams is not used in the conventional microscope since these now have a different wavelength and the imaging system is inappropriate. This leads to a loss of resolution because of chromatic effects. The elastically scattered electrons have the same wavelength as the incident beam and can produce

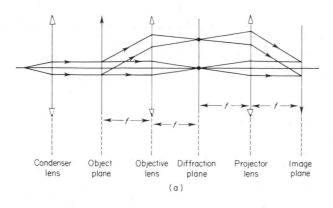

Condenser   Object   Objective Diffraction   Projector   Image
lens   plane   lens   plane   lens   plane

(a)

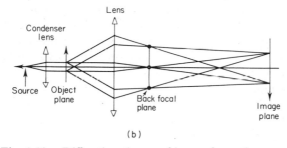

(b)

Fig. 1.10.   Diffraction theory of image formation.

interference effects. Since the electrons used in image formation are scattered without loss of energy the information about the object is impressed on the electron beam in the form of phase information and these electrons would produce no contrast in the Gaussian image plane of a perfect optical system. Contrast must therefore be produced in the image, either by the use of an aperture in the back focal plane of the objective lens or by the method of out-of-focus phase contrast. The theory of contrast is best understood in terms of diffraction theory of image formation, as propounded by Abbe. This is summarized in Fig. 1.10(a), which shows how the formation of an image of unit magnification can be considered as two equivalent stages of diffraction. Stage 1 represents a transformation from the object plane to the diffraction pattern and stage 2 a second transformation from the diffraction pattern to the image plane. The same theory applies to the more usual arrangement of a single lens producing a magnified image as in Fig. 1.10(b). An important feature of the theory is the fact that the information about the object is displayed in the back focal plane of the objective lens in the form of a spectral analysis of the periodicities present in the object and thus allows the possibility of modifying the amplitude or phase of certain selected diffracted beams in order to produce contrast in the image.

The various methods of contrast production fall broadly into two categories:

(a) Detail greater than about 1·5 nm known as scattering contrast. In this method, contrast is produced by removing some of the scattered electrons from the beam by an aperture placed in the back focal plane of the objective lens, so that those parts of the specimen having a high scattering power will appear deficient of electrons in the image.

A specimen of low differential scattering power, such as a replica of a surface, may be shadowed with a heavy metal. In this way, the areas receiving the metal deposit become strongly scattering, while those in 'shadow' are unchanged. The topography of the surface is thus revealed.

In a crystalline specimen having a high degree of order, interference takes place between the diffracted waves so that those beams related to the structure and its orientation to the electron beam through the Bragg angle will reinforce while the others interfere destructively. In order to resolve the structure of a perfectly periodic system, the aperture must admit the zero-order beam and at least one other beam. If only the zero-order beam passes through the aperture, no periodic structure will be resolved. If we have a crystal that contains a defect, then the strain field in the region of the defect will produce local variations in the diffracting conditions and the presence of the defect may be revealed by a contrast effect, while the periodic structure is not resolved. This contrast method is known to the materials scientist as 'diffraction contrast'. A goniometer stage which varies the orientation of the specimen with respect to the incident beam and the facility for tilting the illumination is of great value for this type of work. If the zero-order beam is excluded and only a diffracted beam allowed to enter the aperture, then the condition known as dark-field illumination is obtained, the image produced by the diffracted electrons showing bright against a dark background.

(b) Detail less than 1·5 nm which is produced by phase contrast. For this method, the illuminating beam should be adjusted for maximum coherence and the objective aperture made as large as possible to admit the diffracted beams. The combined effects of spherical aberration and defocussing of the objective lens can then be used to change the phase of the diffracted beams relative to the undiffracted beam and contrast is produced in a manner analogous to that of the Zernicke phase contrast optical microscope.

Unfortunately, the relative phase changes can only be partially controlled by the microscopist so that some detail appears in positive contrast, some in negative contrast, and some is lost completely. The method is inherently capable of very high resolution and is the subject of much research at the present time.

1.6   ELECTRON BEAMS FOR MICROANALYSIS AND SCANNING MICROSCOPY

An electron beam system for microanalysis must produce an electron spot $0.1-1$ μm in size and this normally requires the source to be demagnified about 250 times. The other requirement is a long working distance ($0.5-1$ cm) from the lens to the specimen for mechanical manipulations and so that X-rays may be conveniently taken off. These conditions are generally met by a double lens system similar to the condenser system of the transmission microscope. The first lens of short focal length produces a highly demagnified image of the cross-over, while the second lens, of longer focal length, projects this image on to the specimen with further demagnification.

The control of the aberrations of the lenses is of great importance in producing a small symmetrical spot, and a stigmator is used as in the microscope. The first lens of very short focal length can be made so that it does not contribute to the overall spherical aberration, but the second lens is more critical in design owing to its longer working distance. Chromatic aberration can be made negligibly small by having stabilised supplies for lens currents and accelerating voltage. Details of the instrumental requirements of the electron-optical column in microanalysis are given in Chapter 6.

There are no conventional imaging systems in instruments of the scanning type. The main function of the electron-optical system is to produce a focussed beam of electrons in the plane of the specimen. The parameters that have to be considered are the beam current, the spot size and the beam convergence.

The maximum current density that can be produced in a focussed spot is given by the Langmuir equation

$$j_m = j_0 \frac{eV}{kT} \alpha^2$$

where j_m = maximum current density, j_0 = current density at the cathode,

T = temperature of the cathode, V = accelerating potential, and α = semi-angular aperture of system.

It is convenient to define a quantity called the brightness β to be the current density per unit area per unit solid angle so that

$$\beta = j_0 \frac{eV}{\pi k T} \qquad \text{and} \qquad j_m = \beta \pi \alpha^2$$

Thus the brightness β is a constant for a given optical system (operating at constant V and T). The beam current can be calculated at any image plane (assuming the system to be free of aberrations and diffraction effects). The brightness β for a tungsten hairpin filament is about 5×10^8 A m^{-2} sr^{-1} and for a field emission source about 2×10^{12} A m^{-2} sr^{-1}, both at 25 kV.

From considerations of the signal-to-noise ratio in the image one can calculate a minimum beam current, current which is necessary in order to record the image in a reasonable time. Thus when the working distance and spot size are specified, the equation

$$i = \frac{\pi d^2 \beta \alpha^2}{4} \simeq 2 \cdot 5 \beta d^2 \alpha^2$$

determines the final aperture size.

Of course the minimum spot size is also dependent on the spherical aberration coefficient of the final lens. The optimum size for the angular aperture, taking into account spherical aberration and diffraction effects, has already been calculated to be about 3×10^{-3} rad.

For the recording of electron-channelling patterns the beam convergence may be more important than the spot size and may be reduced by overfocussing the final lens to give a higher angular resolution as described in the section on the action of the condenser lens (see Section 1.5.1).

FURTHER READING

P. Grivet, *Electron Optics*, Vols. I and II. Translated by P. W. Hawkes. Pergamon Press, Oxford, 1972.

P. W. Hawkes, *Electron Optics and Electron Microscopy*. Taylor and Francis, London, 1972.

2

Electron Diffraction

J. A. BELK

The Royal Military College of Science, Shrivenham, UK

2.1 INTRODUCTION

Electron diffraction experiments were first done by Thomson and by Davisson and Germer in the 1920s in order to help verify de Broglie's proposal that a wave motion was always associated with moving particles. Because of the strong scattering of electrons by matter and the ability of electron beams to be focussed and deflected by magnetic and electrostatic fields, electron diffraction techniques have multiplied in number and become very important in the physical analysis of matter.

The interaction of an electron beam with matter may change its direction and its energy. The scattering of an electron beam can be either elastic scattering, which occurs from a regular array of atoms, or inelastic scattering, which occurs with the various components of an atom or crystal. Elastic scattering gives rise to a diffracted beam with essentially the incident energy while inelastic scattering gives a diffuse pattern and a reduction in energy. Both types of scattering lead to different electron diffraction phenomena, which can be used in the study of composition or structure of materials. Because of the powerful scattering effect of electrons with matter all experiments must be performed in a vacuum and the effective specimen thickness must be very small.

Experiments involving the transmission of electron beams through materials have always posed problems of specimen preparation. Other techniques are limited to the study of the few atomic layers nearest the surface.

The earliest experiments marked off the two main categories of electron diffraction study. Thomson founded the high-energy electron diffraction studies while Davisson and Germer used much lower electron energies, so establishing this alternative technique. This latter technique is restricted to surface studies, and requires purpose-built apparatus. High-energy techniques include electron microscopy, both transmission and scanning, electron probe microanalysis, and selected area electron diffraction, all techniques which combine to give a close characterisation of the material under study.

There are two theoretical approaches to the interpretation of electron diffraction results. The first, the *kinematical theory*, assumes that each atom receives the same incident radiation, that is the diffracted beam does not detract from the incident beams energy. This applies where there is no strongly diffracted beam, where the specimen is thin, where the electron energy is high and where the atomic number of the material is low. Other situations require the *dynamical theory*, which accepts that the energy of the diffracted beam reduces the energy of the transmitted beam. It also allows for the fact that the primary diffracted beam is in turn diffracted again along the direction of the transmitted beam; hence, a dynamical equilibrium is established between the transmitted beam and the various diffracted beams. These theories can account not only for the shape of electron diffraction patterns but also for the contrast effects which occur in electron microscopy. An electron diffraction pattern can be formed with or without a series of lenses between the specimen and the photographic plate or other recording device. If there are no lenses the technique is known as general area electron diffraction. A transmission electron microscope has lenses between the specimen and the viewing screen and here it is possible to select a small area of the specimen for examination, either by focussing the incident beam or blanking off the electrons from the part of the primary electron image that is not required. This technique is known as selected area electron diffraction, and it allows very small specimens or very small components of a larger specimen to be studied.

If there are no lenses between the specimen and the viewing screen then the distance between them is L, the camera length. If lenses are used to magnify the diffraction pattern there will be an effective value of L that is larger than the actual physical distance from specimen to screen. Bragg's law states that for diffraction to take place

$$\lambda = 2d_{hkl} \sin \theta$$

where λ = wavelength of radiation, d_{hkl} = spacing of (hkl) diffracting planes, and 2θ = angle between transmitted beam and diffracted beam. θ is always a small angle so we may write $\sin \theta = \theta$. If r_{hkl} is the distance of the

(hkl) diffraction spot from the centre of the pattern we have

$$\lambda = d_{hkl} \times 2\theta = \frac{r_{hkl}}{d_{hkl}}$$

Hence

$$d_{hkl}r_{hkl} = \lambda L \tag{2.1}$$

For large angles and greater accuracy L should be multiplied by $(1 + 3r_{hkl}^2/8L^2)$. λL is known as the camera constant and it is clearly essential to know what its value is with some accuracy before electron diffraction results can be interpreted.

Clearly the value of λL will depend on the accelerating voltage and the currents in the various magnetic lenses, assuming the geometry of the system stays constant. In an electron microscope the value of λL can vary by a few per cent not only between repeat settings of the microscope in the diffraction mode but also with direction around and distance from the centre of the pattern. Some method of standardization is necessary, particularly for the analysis of diffraction patterns from unknown materials. This can be done by introducing a fine-grained, usually evaporated, standard substance, such as graphite, gold or thallium chloride, and recording its diffraction pattern without altering any of the instrumental settings and preferably superimposed on the same photographic plate. When using a thin foil of a known material there is a built-in standard, but values of λL obtained in this way may not be particularly accurate owing to the high intensity and hence large size of the diffraction spots. The formation of Kikuchi lines is covered in Chapter 4 as are contrast theory, dislocation imaging and the weak beam technique.

2.2 THE RECIPROCAL LATTICE AND EWALD SPHERE CONSTRUCTION

A crystal containing a series of diffracting planes distant d_1, d_2, d_3, etc., from the origin can be represented for diffraction purposes as a lattice of points in reciprocal space. The reciprocal lattice consists of points distant $1/d_1, 1/d_2, 1/d_3$, etc., from the origin along directions that are parallel with the normals of the respective planes. A simple example is shown in Fig. 2.1 of the low index diffracting planes and reciprocal lattice for a face-centred cubic crystal. The construction is shown for the (200), (220) and (111) planes but if the (020), (002), (202), (022) and (222) planes are included the reciprocal lattice is seen to be body-centred cubic. A precisely similar

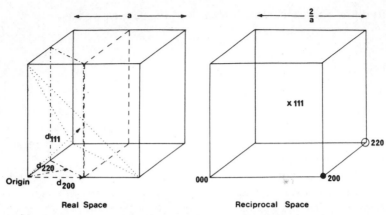

Real Space · **Reciprocal Space**

Fig. 2.1.   Illustration of relationships between diffracting planes in real space and reciprocal lattice points for face-centred cubic crystal.

construction applies for any non-cubic crystal where each diffracting plane in real space is represented by a point in reciprocal space. This concept is convenient in interpreting diffraction patterns because each set of diffracting planes is responsible for one spot in the diffraction pattern and as given in eqn. (2.1) the distance r_{hkl} of the spot from the centre of the pattern is proportional to $1/d_{hkl}$. The value of the reciprocal lattice is seen most clearly in the reflecting sphere or Ewald sphere construction. Diffraction can be represented in real space and in reciprocal space in the ways shown in Fig. 2.2. In real space the angle 2θ between a strong diffracted beam and the incident beam is given by Bragg's law. This is represented in reciprocal space by the vector diagram where the incident radiation is represented by a vector, **k**, and the diffracted radiation is given by the sum of that vector and the reciprocal lattice vector $\mathbf{g}_{hkl}$. Vector **k** has length $1/\lambda$ and direction parallel to the incident radiation, and points to the origin of the reciprocal lattice. Vector $\mathbf{g}_{hkl}$ has length $1/d_{hkl}$ and direction normal to the diffracting planes (hkl). If a sphere of radius $1/\lambda$ centred at the origin of the vector **k** cuts a reciprocal lattice point a strong diffracted beam represented by the vector $\mathbf{k}^1$ is given. It can easily be shown that this construction satisfies Bragg's law:

$$|g_{hkl}| = 2|k| \sin \theta$$

$$\frac{1}{d_{hkl}} = 2\frac{1}{\lambda} \sin \theta$$

$$\lambda = 2d_{hkl} \sin \theta$$

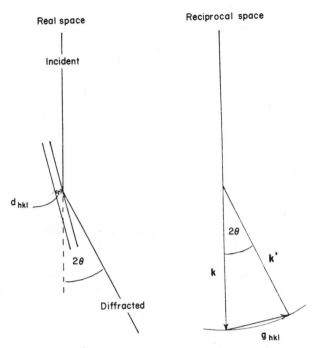

Fig. 2.2. Representation of diffraction in real space and reciprocal space.

The wavelength of high-energy electrons is short, 100 keV electrons having a wavelength of 3·7 pm. The Ewald sphere will therefore have a radius of $270 \times 10^9 \text{ m}^{-1}$. For copper the d_{200} is 181 pm, hence the length of the reciprocal lattice vector $\mathbf{g}_{200}$ is $5·5 \times 10^9 \text{ m}^{-1}$. The spacing of the reciprocal lattice is very much less than the radius of the sphere, hence the centre of a diffraction pattern closely resembles a plane section through the reciprocal lattice perpendicular to the direction of the incident radiation. A more extensive version of the face-centred cubic reciprocal lattice is shown in Fig. 2.3, together with three sections taken on planes perpendicular to the [001], [101] and [111]. Other sections can easily be constructed and these sections have an exact geometric similarity with the diffraction patterns arising from an electron beam incident along the [001], [101] and [111].

Real crystals give reciprocal lattice points of a size inversely proportional to the dimensions of a crystal. A thin sheet crystal, for instance, will give rod-like reciprocal lattice points with the axis of the rod normal to the plane of the sheet. In a similar way rod-like crystals give circular disc-shaped reciprocal lattice points. This means that for thin crystals such as must be

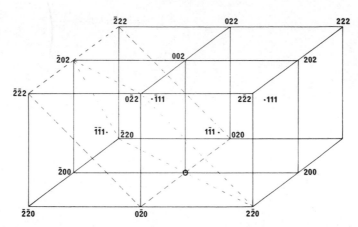

Fig. 2.3.   Part of the reciprocal lattice for a face-centred cubic crystal showing the
[101] and [111] sections.

used in electron diffraction in order to obtain penetration, the Ewald sphere
will cut a large number of reciprocal lattice points owing to their extended
size.

2.3   TO INDEX AN ELECTRON DIFFRACTION PATTERN

Each spot on the pattern occurs at a distance r_{hkl} from the origin. This
distance is related to the spacing d_{hkl} of the diffracting planes by eqn. (2.1):

$$r_{hkl} = \frac{\lambda L}{d_{hkl}}$$

where λL is the camera constant of the instrument used to produce the
pattern. To index a diffraction pattern select three spots that make a
parallelogram with the origin and label them 1, 2 and 3 as shown in Fig. 2.4.
Let them have Miller indices $(h_1\, k_1\, l_1)$, $(h_2\, k_2\, l_2)$ and $(h_3\, k_3\, l_3)$.

Measure r_1, r_2 and r_3 and hence find d_1, d_2 and d_3 from eqn. (2.1). The
values of d determine what type of diffracting plane is responsible for each
of the three spots as long as the specimen is not an unknown material. The
indices of the three spots must be assigned in such a way that

$$h_1 + h_2 = h_3$$
$$k_1 + k_2 = k_3$$
$$l_1 + l_2 = l_3 \tag{2.2}$$

An example will help to clarify the sequence. In the case given, suppose an aluminium crystal gave r_1, r_2 and r_3 as 16·5 mm, 27·4 mm and 36·1 mm in an instrument whose λL is $3·35 \times 10^{-12}\,m^2$.

The diffracting planes in aluminium have d values given below.

$$\{111\}\ 0·2338\ nm$$
$$\{200\}\ 0·2025\ nm$$
$$\{220\}\ 0·1432\ nm$$
$$\{311\}\ 0·1221\ nm$$
$$\{331\}\ 0·0929\ nm$$
$$\{420\}\ 0·0905\ nm$$

$d_1 = 0·203\ nm$ hence spot *1* results from $\{200\}$ diffraction.

$d_2 = 0·122\ nm$ hence spot *2* results from $\{311\}$ diffraction.

$d_3 = 0·0928\ nm$ hence spot *3* results from $\{331\}$ diffraction.

If we designate *1* as the (200) spot a solution which satisfies eqns. (2.2) is that *2* is the (113) spot, and *3* is the (313) spot. There are other self-consistent solutions, the important criterion being that the three sets of indices should obey eqn. (2.2).

If a set of three Miller indices that satisfy eqn. (2.2) can be found they are probably correct and unique, but as a check the angle between $(h_1\,k_1\,l_1)$ and $(h_2\,k_2\,l_2)$ can be computed and this should be the same as the angle subtended by spots *1* and *2* at the centre of the pattern.

The direction **B** along which the electron beam travelled before diffraction can now be determined as it is the zone axis $[u\,v\,w]$ of the diffracting planes. It must be perpendicular to the directions $[h_1\,k_1\,l_1]$, $[h_2\,k_2\,l_2]$ and $[h_3\,k_3\,l_3]$ hence

$$uh_1 + vk_1 + wl_1 = 0, \text{ etc.} \tag{2.3}$$

A pair of such equations can be solved by writing a matrix as follows:

$$h_1 \quad k_1 \quad l_1 \quad h_1 \quad k_1 \quad l_1$$
$$h_2 \quad k_2 \quad l_2 \quad h_2 \quad k_2 \quad l_2$$

Deleting the extreme left- and right-hand columns, u, v and w are then equal to the determinants given by the left two columns, the middle two columns, and the right two columns respectively, that is

$$u = k_1 l_2 - k_2 l_1$$
$$v = l_1 h_2 - l_2 h_1$$
$$w = h_1 k_2 - k_2 h_1$$

It may be necessary to rationalize $[u\,v\,w]$ to minimum whole numbers. Again with the example from above

$$\begin{array}{ccc} 2 & 0020 & 0 \\ \diagdown\!\!\!\!\diagup & & \diagdown\!\!\!\!\diagup \\ 1 & 1311 & 3 \end{array}$$

$u = 0, \ v = 0 - 6 = -6, \ w = 2 - 0.$

The zone axis **B** is

$$[0 \ \bar{3} \ 1]$$

Conventionally the zone axis is considered to point upwards from the crystal. If the spots are designated *1* and *2* as shown in Fig. 2.4, then the formula for u, v and w will give the correct sense for **B**.

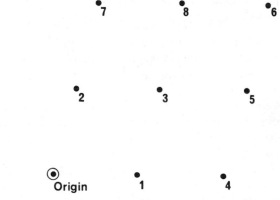

Fig. 2.4.   Schematic representation of electron diffraction pattern.

It is now possible to index every other spot on the pattern by using the property that, moving along the vector r_1 from a known spot to an unknown, adds h_1, k_1 and l_1 to the Miller indices; hence the unknown spot is indexed. Similarly, going along the vector r_2 adds h_2, k_2 and l_2.

In Fig. 2.4 spot *4* is obtained by moving along vector r_1 from spot *1*, hence the indices of spot *4* are (2 + 2, 0 + 0, 0 + 0) or (400). Similarly spot *5* is obtained by moving along vector r_1 from spot *3*, hence the indices of spot *5* are (3 + 2, 1 + 0, 3 + 0) or (513). Spot *5* can also be obtained by moving along vector r_2 from spot *4* hence its indices are (4 + 1, 0 + 1, 0 + 3) or (513). Every other spot can be indexed in the same way.

If the space lattice is known but not the value of λL, the ratios of r_1, r_2 and

r_3 can be compared with the ratios of the various d values to decide on the type of planes giving rise to the diffraction spots. For instance, the ratios of r_2 and r_3 to r_1 are 1·661 and 2·188. If these are compared with the table of possible ratios in the face-centred cubic system (Table 2.1) it is clear that they are d_{200}/d_{311} and d_{200}/d_{331} respectively—hence the same solution as before. If the space lattice is not known, similar tables of d spacings for

TABLE 2.1
Table of Ratios of d Values in Face-centred Cubic Crystals

hkl	$\dfrac{d_{111}}{d_{hkl}}$	$\dfrac{d_{200}}{d_{hkl}}$	$\dfrac{d_{220}}{d_{hkl}}$
111	1		
200	1·1547	1	
220	1·6330	1·4142	1
311	1·9149	1·6583	1·1726
331	2·5166	2·1794	1·5411
420	2·5820	2·2361	1·5811

possible lattices must be consulted to obtain a match, or the d values can be calculated and compared with those of known substances in the ASTM index.

The above procedure is suitable for indexing the centre of electron diffraction patterns but there often occur bands of spots away from the centre, which arise from reciprocal lattice points that are not on the same plane as those giving rise to the central spots but on a parallel plane above it. Equation (2.3) is applicable to the central spots, but this must be modified to $uh + vk + wl = C$ where C is an integer, to account for the spots away from the central region, which are said to belong to a different Laue zone. In a symmetrical diffraction pattern the central region of the pattern, or zero-order Laue zone, is circular and there is a circular band of spots representing the second-order Laue zone where $C = 2$. The radius p of the zero-order Laue zone enables an estimate of the foil thickness to be obtained from the formula

$$p = L\sqrt{\frac{2\lambda}{t}}$$

This formula only applies if the foil is thin. If the foil thickness is greater than the extinction distance ξ_g (see Chapter 4) the formula does not apply.

If the diffraction pattern is not symmetrical the Laue zones take up an elliptical shape. The shape and size of the ellipse allow the angle of tilt from a simple orientation to be determined. The ellipse can usually be considered as a circle and if the distance of the centre of the circle from the centre of the diffraction pattern is R then the angle of tilt ϕ is given by

$$\tan \phi = \frac{R}{L}$$

A more accurate way of determining the angle of tilt is to utilise the position of the Kikuchi lines within the diffraction pattern.

2.4   DOUBLE DIFFRACTION

Double diffraction occurs when a strongly diffracted beam suffers further diffraction during its further course through the crystal. This can give rise to intensity changes in diffraction spots, to the appearance of 'forbidden' spots and satellite spots surrounding the main spots in a diffraction pattern. A beam that has been diffracted by the $(h_1 k_1 l_1)$ planes and is then subsequently diffracted by the $(h_2 k_2 l_2)$ planes will give rise to a spot with indices $(h_1 + h_2,\ k_1 + k_2,\ l_1 + l_2)$. Forbidden spots do not appear in diffraction patterns of face-centred or body-centred crystals. However, they do arise in the patterns from hexagonal close-packed crystals. The (0001) spot can arise by double diffraction from the (01$\bar{1}$1) and (0$\bar{1}$10) spots, for instance.

Small satellite diffraction spots surrounding the main spots can arise in twinned face-centred cubic crystals. They are symmetrically disposed about the main spots with the primary twin diffraction spots. Two-phase alloys, particularly where there is an orientation relationship between the phases, can give rise to striking double diffraction effects. Similar effects are given by specimens covered with an oxide film. Groups of closely spaced spots are associated with each of the matrix spots.

2.5   ZONE AXIS PATTERNS

If the electron beam is incident along a major symmetry axis of the crystal, many sets of planes will contribute to the diffraction and a complex effect will result. The electron diffraction pattern consists of a number of equally bright spots around the central spot. However, the image seen in an electron microscope is of the intersection of a series of bend contours, each of which results from the diffraction of electrons by a particular set of planes. The axis common to all the sets of diffracting planes is the zone axis, hence these patterns are known as zone axis patterns (Fig. 2.5). The patterns have a characteristic star shape and the centre of the patterns contains information which allows more details of the crystal structure to be obtained than is

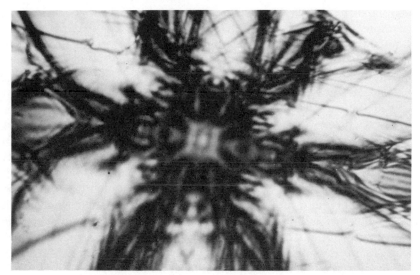

Fig. 2.5.   Zone axis pattern of aluminium [110] at 100 kV.

possible from an electron diffraction pattern. Analysis of this type of information has been called 'real space crystallography'.[1] Zone axis patterns can be obtained either by viewing the image of a bent crystalline foil in an electron microscope under substantially parallel incident illumination or by using a strongly convergent beam and studying the so-called convergent beam patterns. These occur in the plane of the diffraction pattern and contain information very similar to the bend contour zone axis patterns. Some modification of the microscope is required to obtain the convergent beam pattern but there is no necessity to have a suitably bent specimen. The angular resolution obtainable with convergent beam patterns is much greater than with bend contour patterns and it is possible to record the intersection of fine high-order lines within the zone axis pattern whose position is strongly dependent on the electron energy and lattice parameter. They can thus be used for accurate calibration of the accelerating voltage and for studying changes in lattice parameter, caused, for instance, by alloying or thermal expansion. A precision of 2 in 10^4 is possible in lattice parameter determination.[2]

The main attraction of real space crystallography is that by examining a series of zone axis patterns it is often possible to determine the space group. This can be done on a very small specimen, such as a grain in a fine-grained multiphase alloy. Unlike conventional X-ray determinations it does not simply give the average over a whole block of material, hence point-to-point variations of structure can be detected. The zone axis patterns give the symmetry elements of the Laue group of the crystal structure. Dark field images give not only the point group but also screw axes and glide planes, hence the whole space group can be determined from a combination of zone axis patterns and dark field images.

At a particular voltage known as the critical voltage the centre of a pattern is a black spot. Above and below this value of voltage, rings appear in the centre of the pattern. The number and spacing of the rings can be used to determine the thickness of the foil to a few nanometres and the presence of interstitial solutes can change the intensity of some of the rings.

Applications have so far enabled different compositions of stainless steel to be distinguished from their convergent beam patterns and isomorphous compounds such as SnS_2, SnSSe and $SnSe_2$ can be distinguished, although the S and Se atoms simply replace each other in the structure. Some examples of the use of convergent beam diffraction in the study of precipitation phenomena are given in Chapter 3.

REFERENCES

1. J. W. Steeds, *Nature*, London, 1973, **241**, 435.
2. G. M. Rackham, P. M. Jones and J. W. Steeds, *Eighth International Conf. Electron Microscopy*, Australian Academy of Science, Canberra, **1**, p. 336, 1974.

BIBLIOGRAPHY

For a full description of the production, indexing and interpretation of electron patterns see:

P. B. Hirsch, A. Howie, R. B. Nicholson, D. W. Pashley and M. J. Whelan, *Electron Microscopy of Thin Crystals*. Butterworth, London, 1965.
B. E. P. Beeston, *Practical Methods in Electron Microscopy* (ed. A. M. Glauert). Vol. 1, pp. 193–323. North-Holland Publishing Co., Amsterdam, 1972.

3

Precipitation Studies

G. W. LORIMER

University of Manchester, UK

3.1 INTRODUCTION

When *Electron Microscopy and Microanalysis of Metals* (Eds. J. A. Belk and A. L. Davies; Elsevier, London) was published in 1968 the chapter on precipitation studies was concerned with the application and interpretation of transmission electron microscopy (TEM). During the last ten years TEM has continued to be used to study precipitation phenomena: it is now supplemented by high-resolution microscopy, high-voltage electron microscopy (HVEM), analytical electron microscopy and convergent beam techniques. In 1968 electron microscopes with high-resolution capabilities, sophisticated goniometer stages and electronic beam tilting were available in only a few, specialised, research laboratories; now most electron microscopists who are studying crystalline materials have access to such instruments. A similar dissemination of analytical capability—in the form of scanning transmission electron microscope (STEM) attachments to conventional TEMs and energy dispersive X-ray detectors—is taking place at the present time.

In this chapter the application of conventional techniques of TEM to the study of precipitation phenomena is reviewed. The high-resolution techniques of weak-beam and direct-lattice microscopy are described, as is convergent-beam diffraction. The advantages of the use of HVEM in precipitation studies are outlined and the types of problems to which it can be usefully applied are reviewed. Analytical electron microscopy is briefly described and the possible applications of the technique to the study of precipitation are discussed.

3.2   TRANSMISSION ELECTRON MICROSCOPY

If a particle is visible in a thin foil it must either diffract differently from the matrix or produce a distortion of the adjacent matrix; these effects have been termed precipitate contrast and matrix contrast, respectively.[1] In both the contrast can be interpreted, at least qualitatively, using the 'column approximation', in which the scattering events which occur in the crystal are assumed to be localized to columns approximately 20 Å in diameter (Chapter 4).

3.2.1 Matrix Strain Field Contrast

A precipitate that is coherent but has a different lattice parameter from the matrix will produce an elastic distortion in the adjacent matrix. For a spherical particle in an infinite, isotropic matrix the displacement in the matrix, R, at a distance r is given by

$$R = \frac{\varepsilon r_0^3}{r^2} \tag{3.1}$$

where r_0 is the radius of the particle and ε the *in situ* misfit. The lattice displacements near such a particle are shown schematically in Fig. 3.1(a). A column of crystal adjacent to the particle is distorted and will diffract

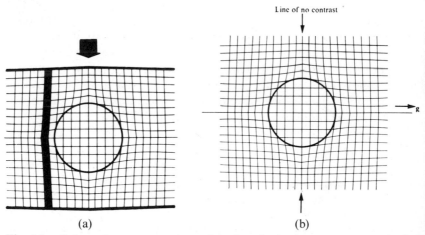

(a)	(b)

Fig. 3.1.   Schematic representation of lattice displacements near a spherical precipitate. (a) Electron beam vertical showing distortion of a column of crystal adjacent to the precipitate. (b) Looking 'down the electron beam' showing the 'line of no contrast' for the operating reflection **g**.[2]

differently from an undistorted column some distance away. This is the
origin of the strain field contrast which is observed in the micrograph (Fig.
3.2). Figure 3.1(b) is a view of the particle in Fig. 3.1(a) looking down the
electron beam and the crystal is orientated so that the set of planes running
vertically in the micrograph are satisfying the Bragg law, i.e. $\mathbf{g}$ is the
operating reflection. The planes that pass through the centre of the particle

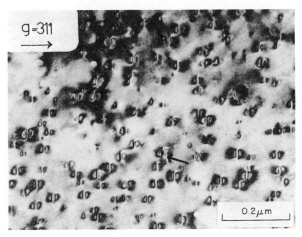

Fig. 3.2. Bright-field electron micrograph showing matrix strain-field contrast
from spherical Co particles in a Cu–Co alloy. The 'line of no contrast' is
perpendicular to $\mathbf{g}$. An 'anomalous image' from a particle near the surface is
arrowed.[3]

are undistorted, the product $\mathbf{g} \cdot \mathbf{R} = 0$, and the image will not show strain-
field contrast. This 'line of no contrast' is perpendicular to the operating
reflection $\mathbf{g}$, if the particle has a spherically symmetric strain field, and will
change direction as $\mathbf{g}$ is varied (Fig. 3.2).

The image of the particle is strongly dependent on the value of the
parameter $\varepsilon g r_0^3/\xi_g^2$ where ξ_g is the extinction distance.[3] For values less than
0.2, images of particles are similar to Frank dislocation loops: they consist
of black dots with black and white lobes in both bright field and dark field
and the image is a sensitive function of the position of the particle in the foil.
If $\varepsilon g r_0^3/\xi_g^2$ is greater than 0.2, all precipitates, except those within an
extinction distance of the surface, exhibit a pair of symmetrical dark lobes
in a bright-field image (or bright lobes in a dark-field image) with an obvious
line of no contrast (Fig. 3.2).

Particles close to the surface exhibit large asymmetrical images due to the surface relaxation of the strain field. The images are identical in bright field and dark field for precipitates at the top of the foil, due to absorption, and complementary for particles at the bottom of the foil. These 'anomalous images' (one such image is indicated in Fig. 3.2) can be used to determine the sense of the misfit ε: if the atomic volume of the precipitate is larger than the matrix (ε positive), the lobe on the side of the precipitate in the direction of

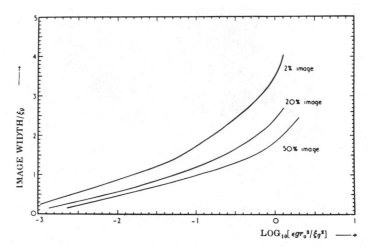

Fig. 3.3.   Predicted variation of the image width as a function of the parameter $\varepsilon \mathbf{g} r_0^3 / \xi \mathbf{g}^2$.[3]

positive $\mathbf{g}$ will be dark in a dark-field image; if ε is negative then, under the same diffracting conditions, the corresponding lobe will be light.

If the precipitates are sufficiently large that $\varepsilon g r_0^3 / \xi_g^2 > 0 \cdot 2$ it is possible to determine the size, as well as the sense, of the misfit ε by examining the width of the strain-field image.[3] For such a precipitate the width of the image is almost independent of its position in the foil, as long as it is not within an extinction distance of the surface. The 2 %, 20 % and 50 % image width of the strain is measured (the detailed procedure is given by several authors[1,3,4]) under two beam conditions with $\mathbf{s} = 0$. The precipitate radius, r_0, is measured and $\mathbf{g}$ and ξ_g determined. The measured image width is related to a value of $\varepsilon g r_0^3 / \xi_g^2$ using the calibration curves of Ashby and Brown,[3] which are reproduced in Fig. 3.3, and ε can then be calculated. Considerable care is required if accurate values of the misfit are to be obtained. The major errors arise in the measurement of the precipitate radius and image widths, shape effects from non-spherical precipitates, and matrix elastic anisotropy.

3.2.2 Matrix Displacement Fringe Contrast

Thin planar precipitates with a misfit **R** perpendicular to the plane of the precipitate displace the matrix planes on either side from their normal position. This displacement introduces an additional phase change in the diffracted beam $\alpha = 2\pi\mathbf{g} \cdot \mathbf{R}$. Although the matrix is not faulted the image that arises from the displacement is similar to that from a stacking fault, i.e. it has the characteristics of an α boundary: the image consists of a series of bright and dark fringes which are symmetrical in bright field and asymmetrical in dark field. The fringes denote contours of constant depth in the foil and run parallel to the line of intersection of the plane containing

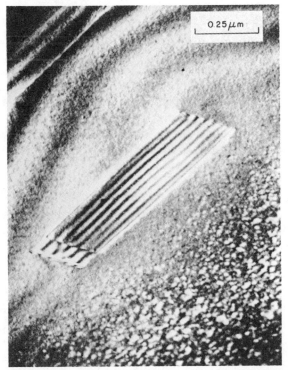

Fig. 3.4.   Bright-field electron micrograph of matrix displacement fringe contrast from a Ca-rich precipitate of augite in an orthopyroxene.[5] The fringes are parallel to the line of intersection of the surface and the precipitate.

the precipitate and the foil surface. The depth periodicity of the fringes is equal to $\xi_g/2$ at $s = 0$ and equals the effective extinction distance $\xi_g^\omega = \xi_g(1 + \omega^2)^{1/2}$ when $\omega \geq 1$ ($\omega = s/\xi_g$). The fringes are most easily seen when the matrix is diffracting strongly and the precipitate diffracting weakly (Fig. 3.4) (s is defined in Chapter 4).

The sense of **R**, whether the precipitate has a larger or smaller atomic volume than the matrix, can be determined by carrying out the same analysis that is used to determine whether a stacking fault is intrinsic or extrinsic (Chapter 4), and the size of **R** can be determined by applying the criterion that the fringe contrast is absent when $\mathbf{g} \cdot \mathbf{R} = 0$.

3.2.3 Moiré Fringe Contrast

Moiré fringe contrast arises from two superimposed lattices of slightly different spacing and/or orientation. In the general case, the moiré spacing D is given by

$$D = \frac{d_1 d_2}{(d_1^2 - d_2^2 - d_1 d_2 \cos \phi)^{1/2}} \tag{3.2}$$

where d_1 and d_2 are the relevant lattice planes of the matrix and precipitate and ϕ is the angle of rotation. For the 'parallel' moiré $D = d_1 d_2/(d_1 - d_2)$ ($\phi = 0$) and for a pure 'rotation' moiré $D = d/\phi$. The most common moiré encountered in precipitation studies is the 'parallel' one. The pattern will be observed only when the relevant matrix and precipitate reflections are both excited and, for the 'parallel' moiré, the fringe pattern will be perpendicular to **g**. The spacing D is independent of the value of **s**.

3.2.4 Precipitate Structure Factor Contrast

The structure factor F_g contains information concerning the atomic species and their position in the unit cell,

$$F_g = \sum_n f_n \exp(2\pi i \mathbf{g} \cdot \mathbf{r}_n) \tag{3.3}$$

where f_n is the atomic scattering factor and $\mathbf{r}_n$ the position vector of the nth atom in the unit cell. A variation in F_g produces a change in the extinction distance as $\xi_g \propto 1/F_g$. Thus the extinction distance in a column of crystal containing a precipitate will be different from a column that passes through the matrix alone, with the result that the thickness fringes at the edge of a foil will be displaced in the columns that contain precipitates. The contrast mechanism is useful for imaging small, coherent precipitates that have

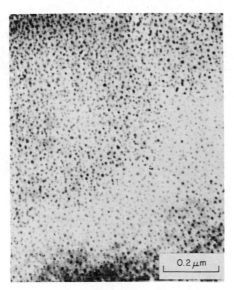

Fig. 3.5.   Bright-field electron micrograph of GP zones in an Al–Ag alloy imaged
by precipitate structure factor contrast.[6]

lattice parameters similar to the matrix and do not produce matrix strain or
give rise to orientation contrast, e.g. Guinier Preston (GP) zones. Figure
3.5 shows GP zones in an Al–Ag alloy which have been imaged using
structure factor contrast.

3.2.5 Precipitate Orientation Contrast and Orientation Relationships
Orientation contrast is the most common type of precipitate contrast and
occurs when the precipitate and adjacent regions of the matrix are
diffracting with different intensities. It can only arise when the precipitate
has a sufficiently different crystal structure from the matrix that it produces
its own diffraction pattern. The most effective use of orientation contrast is
obtained when a high-resolution dark-field image is formed with a
precipitate reflection: the precipitate appears white on a dark background
(Fig. 3.6). This image can be used to make accurate measurements of
precipitate shape and size distribution as the image is free of matrix strain
effects which would make accurate measurements of the shape and size of
the precipitates impossible in a bright-field image.

If the precipitate is coherent or semi-coherent with the matrix (or has
been so at the early stages of growth) it will have an orientation relationship
with the matrix: certain planes and directions in the precipitate will be

Fig. 3.6. Dark-field HVEM (1000 kV) electron micrograph of a natural alkali feldspar formed with a precipitate reflection from the albite ($NaAlSi_3O_8$) phase. The fine striations within the precipitate are caused by twinning.[7]

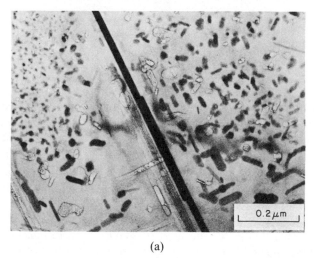

(a)

Fig. 3.7. Al–Zn–Mg alloy aged to contain large η precipitates. (a) Bright-field micrograph in which the large η lath is diffracting strongly and showing orientation contrast. (b) Electron diffraction pattern from (a) showing both precipitate and matrix diffraction spots. (c) Indexed diffraction pattern shown in (b). (d) Stereographic projection on the η precipitate superimposed on a projection of the Al alloy matrix showing the orientation relationship.[8]

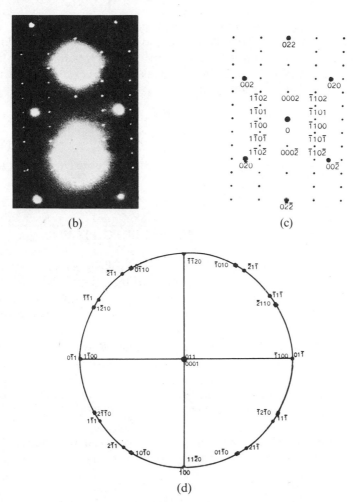

(b)

(c)

(d)

Fig. 3.7—contd.

parallel to specific planes and directions in the precipitate. To determine the orientation relationship a diffraction pattern is obtained that contains matrix and precipitate spots and both patterns are indexed. Figure 3.7(a) is a bright-field micrograph of an η (MgZn$_2$) precipitate formed in an Al–Zn–Mg alloy and Fig. 3.7(b) is the associated diffraction pattern. The indexed diffraction pattern is shown in Fig. 3.7(c). The information on the indexed diffraction pattern is transferred to a composite stereographic projection: a stereographic projection of both the matrix and the precipitate, in the correct relative orientations as obtained from the diffraction pattern (Fig. 3.7(d)). Two or three diffraction patterns should be obtained that contain both matrix and precipitate spots to confirm the orientation relationship between the two phases. In the example shown in Fig. 3.7 the orientation relationship is

$$(0001)_{MgZn_2} \parallel (110)_{matrix}$$

$$[11\bar{2}0]_{MgZn_2} \parallel [100]_{matrix}$$

By convention, orientation relationships are usually quoted in terms of one set of parallel planes and one parallel direction.

3.3 COUNTING AND SIZING PRECIPITATES

The average precipitate diameter $\bar{D}$ (or the mean caliper diameter for non-spherical particles), the volume fraction V_f and the size distribution $N(D)$ are three characteristics of precipitate distribution that are of interest to the microscopist. Once these have been determined, other parameters, such as surface area of particles per unit volume, can be calculated.

The simplest method for determining $\bar{D}$, V_f and $N(D)$ is to take a direct replica from a polished and etched surface *which faithfully reproduces a planar section through the sample.* If such a replica can be prepared, the procedures that have been developed for quantitative optical microscopy can be applied (see, for example, the text edited by DeHoff and Rhines[9] or the review by Pickering[10]).

The V_f of the precipitate can be obtained from a micrograph of a surface replica by the techniques of lineal analysis, areal analysis or point counting, for

$$V_f = A_f = L_f = P_f \tag{3.4}$$

where A_f is the area fraction, L_f is the linear fraction (as measured from the proportion of randomly distributed line segments in the precipitate), and P_f is the point fraction (as determined from the proportion of randomly distributed points in the precipitate).

In order to determine the true size distribution $N(D)$ it is necessary to correct the observed size distribution $\mu(d)$ for the effects of sectioning: precipitates with centres above or below the plane of section will appear as circles of intersections smaller than their true diameter and lead to an apparent increase in the proportion of small particles. The correction procedures were first developed by Scheil[11] for spherical particles and have been extended to plate-shaped particles[12] and problate and oblate spheroids.[13]

If $N(D).dD$ is the number of spheres per unit volume with diameters in the range $D \pm dD/2$, the observed diameter distribution of circles intersected by a plane is $\mu(d)$ where

$$\mu(d) = \int_{D=d}^{\infty} d(D^2 - d^2)^{-1/2} N(D).dD \qquad (3.5)$$

To evaluate $N(D)$ the integral is approximated by a series summation: the observed size distribution $\mu(d)$ is divided into a series of equal intervals, usually 15. The number of particles per unit volume is calculated for each size group, starting with the largest and proceeding systematically to the smallest. A contribution to each size group can arise from all of the groups of larger particles and these must be subtracted from the observed number. (Underwood[14] has pointed out that in the Scheil analysis each calculation of the true number of particles in any size group depends upon all of the calculations for the larger size groups that have been carried out previously. Thus errors persist throughout the calculation and become larger as the particle size decreases. Improvements to the methodology of Scheil have been developed that use the measured values of circles of intersection at each stage of the calculation. The relative merits of these techniques are discussed and the tables used to apply them are given in the article by Underwood.) Once the true size distribution $N(D)$ has been obtained the average particle diameter, $\bar{D}$, and V_f can be calculated. The direct-replica technique is particularly useful for particles that are too small to resolve optically and too large to be consistently retained in a thin foil, e.g. $\bar{D} \approx 0 \cdot 1 - 5 \, \mu m$.

Extraction replicas can be used to determine $\bar{D}$ and $N(D)$ without applying a Scheil-type analysis to correct for the effects of sectioning. As the etching time is increased the maximum size of particle that can be successfully extracted will also increase. If the particle size distributions $N(D)$ are plotted as the etching time is increased these will stabilize when the etch is deep enough to extract the largest particles and $\bar{D}$ can be calculated from the observed size distribution. V_f cannot be obtained by this technique as the etched surface is not a planar section. The extraction replica technique works well for etchant-resistant precipitates such as oxides and carbides when the volume fraction of the precipitates' phase is small, i.e. sufficiently low that 'clumping' and overlap of the extracted precipitates do not occur. When there are small precipitates present—precipitates with diameters less than 10 nm—it is important to check the efficiency of the extraction technique by shadowing the etched surface *before* carbon is evaporated on to it.

When the average precipitate diameter is less than 0·1 μm the usual practice is to use thin foil specimens. The quantitative analysis of precipitate distributions in thin foils is based on the assumption that the foil is bounded by plane surfaces. In reality this is often not the true situation and there may be preferential attack of either the precipitate or the matrix. The effects are small for particles which are much smaller than the foil thickness t; however, when $\bar{D}/t \approx 1$ they can be significant. In a micrograph of a thin foil corrections must be made to the observed size distribution of particles for the effects of sectioning, at the top and bottom surfaces, and particle overlap. The procedures outlined below for making these corrections are those given by Hilliard.[15] For spherical particles the observed distribution $N(d)$ is given by

$$N(d) = tN(D)_{D=d} + \mu(d) - M(d) \qquad (3.6)$$

where t is the foil thickness, $\mu(d)$ the distribution of circle diameters identical to that appearing on a two-dimensional plane through the structure, and $M(d)$ projections lost due to overlap.

If eqn. (6) is integrated over d the number of images observed, N_A, is given by

$$N_A = N_V(t + \bar{D}) - M_A \qquad (3.7)$$

In the absence of overlap the number of particles per unit area N_A counted on an electron micrograph is given by

$$N_A = N_V t + N_V \bar{d} \qquad (3.8)$$

where $\bar{d}$ is the average particle diameter. The second term arises from particles with centres outside the foil within $\bar{d}/2$ of the top and bottom surfaces. If N_A is measured as a function of foil thickness, N_V can be obtained from the gradient of a plot of N_A versus t. To obtain $N(D)$ it is necessary to carry out a Scheil-type of analysis similar to that described earlier for the surface replica.

If the sample is thick compared with the average sphere diameter $\mu(d)$ is small and can be estimated as follows. A test line is drawn at random on the projected image and the distribution $N(L)$ measured, where $N(L) . dL$ is the number of chords of length $L \pm dL/2$ formed by the intersection of a unit length of test line with the circular images. $N(L)$ bears the same relation to $N(d)$ as $\mu(d)$ does to $N(D)$:

$$\mu(d) = [N_V N(L)_{L=d} / N_A] \tag{3.9}$$

$$\approx (1/t) N(L)_{L=d} \tag{3.10}$$

The overlap correction M_A can be ignored if the volume fraction of the precipitates is small. For example, if an error of 5% is acceptable in the calculation of N_V a correction for overlap is not necessary if $V_f(t/\bar{D}) < 0.04$. If it is necessary to correct for overlap then the analytical procedure given by Hilliard[15] may be used or $M(d)$ can be experimentally determined by examining two or three micrographs of the same area with the sample tilted a few degrees between successive exposures. Once $N(D)$ has been determined V_f, $\bar{D}$ and N_V can be calculated using the following equations:

$$V_f = \frac{\pi}{6} \sum_D D^3 N(D) \Delta D \tag{3.11}$$

$$N_V = \sum_D N(D) \Delta D \tag{3.12}$$

and

$$\bar{D} = \frac{1}{N_V} \cdot \sum_D D \cdot N(D) \Delta D \tag{3.13}$$

3.4 THICKNESS MEASUREMENTS

A number of techniques have been developed to determine the thickness of electropolished or ion-thinned specimens. The 'traditional' methods (see, for example Ref. 1, p. 415), which can be applied to specimens where the thickness varies significantly over a distance of a few tens of micrometres, include the use of thickness fringes, trace analysis and parallax measurements. The periodicity of extinction contours is at a maximum at the exact Bragg conditions, $s = 0$, and equals the extinction distance ξ_g. In a bright-field image light fringes occur at integral values of t/ξ_g. Values of ξ_g can be calculated if the structure factor F_g is known (see Chapter 4) or, for some common materials, can be obtained from tables (see, for example, Ref. 1, p. 495, and Ref. 4, Vol. 3, p. 105). Measurements must be made at

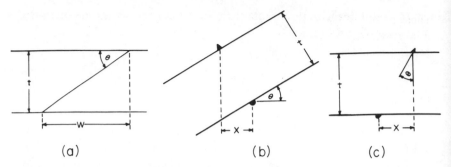

Fig. 3.8.   Geometry of a thin foil for the measurement of specimen thickness from (a) a trace of projected width W; (b) contamination spots formed on a tilted foil; and (c) contamination spots formed on an untilted foil.

the exact Bragg condition as the observed extinction distance ξ_g^ω is a function of the deviation parameter ω. The measurement of extinction fringes can be used to determine specimen thicknesses to an accuracy of ± 15–20% if $t > 3\xi_g$.

If a planar or linear feature, e.g. a slip trace, precipitate or stacking fault, cuts the top and bottom surfaces of the foil, its projected width, W, can be measured (Fig. 3.8(a)) and the thickness calculated as

$$t = W/M \tan \theta \tag{3.14}$$

where θ is the angle between the feature and the foil surface and M the magnification. The major error in all of the trace techniques is the tilt of the foil surface: a tilt of the foil of $5°$ from the horizontal can lead to an error of 5–10% in the calculated thickness, while a tilt of $15°$ can produce an error of almost 50%.[1]

A technique similar to trace analysis for the measurement of foil thickness involves using the parallax separation of objects on the top and bottom of the foil. Hirsch et al.[1] suggest that a small amount of gold evaporated on both surfaces will provide suitable particles for parallax measurements, carried out by stereo microscopy on micrographs that have been tilted a few degrees between exposures (see also Chapter 4). If the transmission electron microscope is fitted with a STEM attachment, or an equivalent probe-forming facility, additional techniques can be used to determine the thickness of specimens. When a probe a few nanometres or tens of nanometres in diameter is focussed on the specimen a contamination spot will be formed on the top and bottom surfaces of the foil under all but the most stringent conditions of specimen cleanliness and microscope vacuum. If the foil is then tilted through an angle θ the separation of the contamination spots X can be used to calculate the thickness[15] using the relationship

$$t = X/M \sin \theta \tag{3.15}$$

as shown in Fig. 3.8(b).

It may be convenient during the microanalysis of samples to place the contamination spots on a tilted foil and make the measurement with the sample horizontal; in this case (Fig. 3.8(c))

$$t = X/M \tan \theta \tag{3.16}$$

The accuracy of the technique depends upon the size of the contamination spot: with probe sizes < 20 nm accuracies of ± 5–10% have been obtained.[17]

The dynamical theory of contrast[1] predicts that subsidiary *maxima* will be observed in bright-field images of extinction contours of integral values of n in the equation

$$\left(s^2 + \frac{1}{\xi_g^2} \right) t^2 = n^2 \tag{3.17}$$

In the kinematical region $s^2 \gg 1/\xi_g^2$ the fringes are equally spaced, if the foil is uniformly bent, and it is easy to calculate t (Ref. 1, p. 510). The reciprocal lattice equivalent of this phenomenon is observed in convergent beam diffraction patterns where the high angle of convergence of the incident probe results in large diffraction discs that reflect the intensity distribution along the reciprocal lattice spikes. The fine scale intensity oscillations formed in convergent beam diffraction patterns can also be used to obtain measurements of foil thickness. These *minima* in the intensity oscillations are related to the foil thickness by eqn. (3.17). The technique has been described by Amelinckx[18] and Kelly et al.[19] The latter claim an accuracy of $\pm 2\%$ in measurements of the thickness of copper foils. This technique requires precise measurement of the fringe spacings (e.g. with a measuring magnifier) and the operator must carry out some calculations.

If the instrument is fitted with an X-ray detector, as well as a probe-forming lens, the intensity of white (background) or characteristic radiation can be used to determine the specimen thickness as both are proportional to thickness for thin specimens. Calibration curves must be determined for each material examined, since the X-ray intensity is a function of density and atomic number, as well as thickness, and background scattering from the specimen holder, grid bars, etc., must be eliminated or subtracted from the true signal. In crystalline materials there is enhanced X-ray production at the exact Bragg condition[20,21] and X-ray intensity measurements must be made well away from extinction contours if they are to be used to monitor the specimen thickness.

An additional complication that arises during the measurement of the thickness of thin foils is the presence of an amorphous surface layer which may be several nanometres thick. These surface films have been detected on thin foils of several aluminium[22,23] and nickel[24] alloys and silicon-rich layers have been detected on electropolished silicon steels in the author's laboratory. The films invariably have a different composition from the matrix, which complicates thin foil analysis, and they will also affect thickness measurements. If an amorphous film is formed on the surface then a technique that determines the thickness of crystalline materials, e.g. convergent-beam diffraction or the projected width of a precipitate or length of dislocation line, must be employed rather than one that measures the total thickness, e.g. contamination spots, projected width of a dislocation slip trace, or white X-ray intensity.

3.5 WEAK-BEAM MICROSCOPY

In weak-beam microscopy, described in more detail in Chapter 4, a high-resolution, dark-field image is formed with a matrix reflection that is weakly excited; s is large.[25,26] It is an extension of conventional (strong-beam) matrix strain field contrast and is used where a high-resolution image is required. At the Bragg condition, $s = 0$, the image width formed by strain-field contrast is large (see, for example, Fig. 3.2); as s is increased the image width decreases and at large values of s it approaches the true width of the precipitate. Weak-beam microscopy can be used as a standard technique to decrease the observed image widths of precipitates, when viewed using matrix strain contrast. However, as the deviation from the Bragg condition is increased, long exposures are necessary and the specimen stage must be very stable and the instrument effectively insulated from vibrations in order to obtain high-resolution micrographs.

Figure 3.9(a) is a conventional bright-field micrograph of an Al–Cu alloy aged to contain θ'' precipitates which are imaged by matrix strain contrast, while Fig. 3.9(b) is the weak-beam image of the same area. Figure 3.9(c) shows microdensitometer traces from three individual precipitates. From these results, and similar ones on alloys aged to contain GP zones, Yoshida et al.[27] were able to construct models for the structure of single-layer GP zones and multilayer θ'' precipitates. In such investigations it is

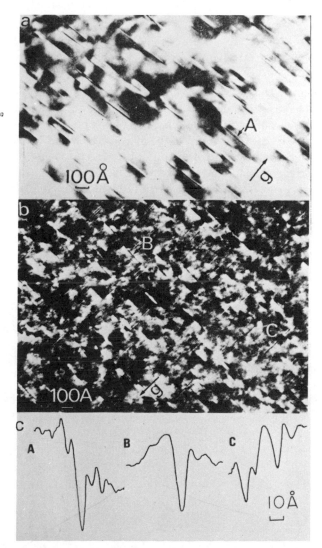

Fig. 3.9.   Bright-field and dark-field micrographs of θ'' precipitates in an Al–Cu alloy. (a) Bright-field image showing matrix strain field contrast. (b) Dark-field weak-beam image formed with the 020 matrix reflection and $s \approx 0.9 \times 10^{-2}\,\text{Å}^{-1}$.
(c) Micro-photometer traces across the images indicated in (a) and (b).[27]

necessary to compare the observed images (Fig. 3.9(c)) with calculated values.

3.6   DIRECT LATTICE MICROSCOPY

If the transmitted beam and one diffracted beam from a *very* thin specimen are allowed to pass through the objective aperture of the microscope, and if the objective lens has a sufficiently high resolving power, a periodic fringe pattern will be formed as a result of the phase interference between the two beams. Under carefully controlled conditions the periodicity of the fringes in the image corresponds to the spacing of lattice planes in the specimen. Figure 3.10 is a direct lattice image of the (100) planes in an orthopyroxene (a chain silicate with $a \sim 1.83$ nm, $b \sim 0.89$ nm, and $c \sim 0.52$ nm), in which GP zones one lattice-parameter thick are resolved. The GP zones are plate-shaped and extend through the specimen so that, in this case, the interpretation of the image is fairly straightforward.

When the specimen thickness is several times the precipitate diameter

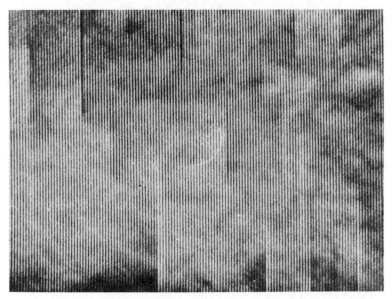

Fig. 3.10. Direct lattice micrograph of an orthopyroxene ($a = 1.83$ nm, $b = 0.89$ nm, $c = 0.52$ nm) showing GP zones one lattice-parameter thick. The lattice fringe spacing is 1.83 nm.[28]

and the direct lattice images show significant deviations from periodicity a one-to-one correspondence between the specimen and the image can no longer be assumed. The phase contrast image, a projection of the lattice potential, is a sensitive function of the specimen thickness t, the spherical aberration coefficient C_s and the amount of defocus Δf of the objective lens. Direct lattice images should be compared with computed images (at specified values of C_s, t and Δf) if quantitative measurements of precipitate structure or matrix strain are to be made.

3.7   HIGH-VOLTAGE ELECTRON MICROSCOPY (HVEM)

The main advantage of HVEM (~ 1000 kV) as compared with conventional (~ 100 kV) TEM is the increase in the thickness of the specimen from which quantitative image or diffraction information can be obtained. The variation in penetration with voltage is almost linear for low density materials but falls off to approximately a thickness versus (voltage)^2 relationship for the transition elements (Fig. 3.11). The ability to examine thick specimens is particularly useful when it is desired to determine the distribution of phases on the 1 μm to 0·1 mm scale and to correlate this with high resolution, 10 nm to 0·1 μm, detail (e.g. Fig. 3.6).

When specimens are prepared by ion beam etching, the standard technique for making thin foils from minerals, ceramics and 'as-cast' metallic specimens, a 'hill-and-valley' topography is developed during the

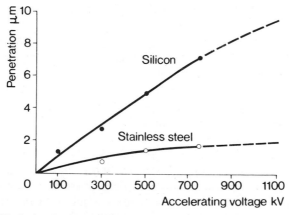

Fig. 3.11.   Variation in penetration with accelerating voltage for silicon ($Z = 14$) and stainless steel ($Z \approx 26$).[30]

thinning process. Although the 'valleys' are usually transparent to 100 kV electrons, higher voltages are required if the 'hills' are to be penetrated as well. Oxides, sulphides and carbides in commercial steels are often too thick to be studied by 100 kV microscopy and the HVEM is an effective method of obtaining crystallographic and morphological information from such particles.[29-31] When a sample contains a high density of small particles, such as a precipitation hardening alloy aged to peak hardness, the HVEM

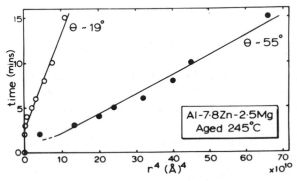

Fig. 3.12.   Average growth rates of precipitates on two grain boundaries in an Al–Zn–Mg alloy measured *in situ* in the HVEM.[32]

offers little to the microscopist: the overlap of image detail in thick specimens is a distinct disadvantage.

At 100 kV the specimen is the thin filling in a sandwich bounded by two highly reactive surfaces and it is usually difficult to carry out *in situ* reactions in the TEM as surface precipitation predominates. At 1000 kV, foils can be sufficiently thick that many precipitation reactions proceed, at least qualitatively, in a similar manner to reactions in bulk specimens. Figure 3.12 is from the work of Butler and Swann[32] and shows the variation of precipitate radius with ageing time on two grain boundaries of an Al–Zn–Mg alloy. It was possible to characterise the misorientations of the two boundaries before the ageing treatment was started and to monitor the rate of growth of the grain boundary precipitates as the reaction proceeded.

Surface oxidation is often a problem when carrying out *in situ* precipitate reactions in the HVEM, although it can be at least partially ameliorated by surrounding the sample with a suitable atmosphere. It is feasible to use an environmental cell in the HVEM as there is enough space between the polepieces of the objective lens to fit the cell, and the electron beam has sufficient energy to penetrate both the atmosphere and the specimen.[33]

TABLE 3.1
*Threshold voltages for displacement
damage (from Ref. 34)*

Element	Threshold voltage (kV)
Al	248
Fe	444
Cu	494
Mo	664
Au	1 092

At voltages above a threshold value the incident electrons will displace atoms from their positions on the lattice (Table 3.1). Typical electron fluxes in the HVEM are sufficiently high to cause severe radiation damage at voltages above this critical value and *in situ* precipitation reactions must be carried out at lower voltages unless it is desired to monitor the effects of electron-produced radiation damage on precipitation.

3.8 CONVERGENT-BEAM DIFFRACTION

The first convergent-beam diffraction patterns were taken by Kossel and Möllenstedt,[35] but the technique which is briefly described in Chapter 2, has not been widely used until recently because standard commercial electron microscopes were not designed to form a highly convergent beam on the specimen. This limitation has been overcome by the recent development of STEM attachments and the improvements to the vacuum in the microscope (the detail in convergent-beam diffraction patterns disappears quickly if the specimen contaminates). In most microscopes the convergent-beam diffraction pattern is formed by focussing a fine probe on the specimen, using the second condenser and objective lenses (Fig. 3.13), and the angle of convergence of the probe (divergence of the diffracted beam), which is set by the size of the condenser aperture (as well as the condenser/objective lens excitation), is usually chosen so that the spots in the convergent-beam pattern just touch or slightly overlap (Fig. 3.14). Convergent-beam diffraction patterns can be obtained at any orientation of the specimen, but information concerning crystal structure (in particular the point group and space group symmetry) is most easily extracted from a zone axis convergent-beam pattern, i.e. when the electron beam is directed along a major symmetry axis of the crystal.

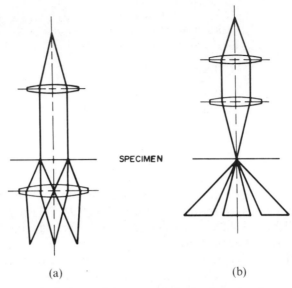

(a)   (b)

Fig. 3.13. Schematic diagram of the ray paths in the electron micrograph to form (a) a 'conventional' electron diffraction pattern, and (b) a convergent beam diffraction pattern.

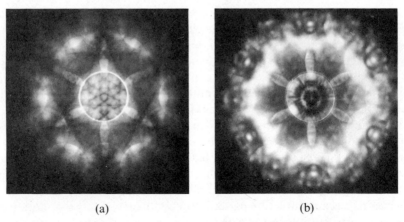

(a)   (b)

Fig. 3.14. $\langle 111 \rangle$ Zone axis convergent-beam diffraction patterns of (a) $M_{23}C_6$, and (b) M_6C. (Thick crystals, $200\,\mu m$ C_2 aperture, $100\,kV$.) The two patterns are distinctly different while the conventional electron diffraction patterns would be identical. Courtesy L. P. Stoter and N. S. Evans.

When the microscope is operated in the 'conventional' diffraction mode (Fig. 3.13), the intensities of the diffracted beams are a sensitive function of specimen thickness and orientation, as well as the crystal structure, and it is difficult to differentiate between electron diffraction patterns from similar structures, e.g. $M_{23}C_6$ and M_6C carbides. Convergent-beam patterns reflect small differences in crystal structure or lattice parameter and have been used to identify precipitates and to measure small differences in lattice parameters between coherent precipitates and the matrix.[36,37] Figure 3.14 shows [111] zone acis convergent-beam diffraction patterns from $M_{23}C_6$ and M_6C carbides.

3.9 ANALYTICAL ELECTRON MICROSCOPY

The technique of analytical electron microscopy, in which X-rays generated in a thin specimen by the incident electron beam are intercepted with a wavelength or energy-dispersive detector, can be used to obtain chemical analyses from regions a few tens of nanometres in diameter. When combined with the high resolution image and crystallographic information that can be obtained in the TEM the technique is a powerful tool for the identification of precipitates and the monitoring of the distribution of solute during phase transformations.

For a thin specimen X-ray absorption and fluorescence can, to a first approximation, be ignored and the ratio of two characteristic X-ray intensities, I_A/I_B, related to the corresponding weight fraction ratio, C_A/C_B, by the equation

$$C_A/C_B = k_{AB} \cdot I_A/I_B \qquad (3.18)$$

where k_{AB} is a constant at a given accelerating voltage which is independent of specimen thickness or composition. A normalisation procedure, e.g. $\sum C_n = 1$, must be used to convert the weight fraction ratios into weight percentages. k_{AB} values can be calculated[38] or determined experimentally[39] and Fig. 3.15 is a comparison between calculated and experimentally determined k_{xSi} values. A full discussion of quantitative X-ray microanalysis is given in Chapter 7.

X-ray absorption in the specimen becomes an important parameter when considering X-rays from light elements, $Z < 14$ (Si), and/or thick samples. To a close approximation absorption can be accounted for by assuming

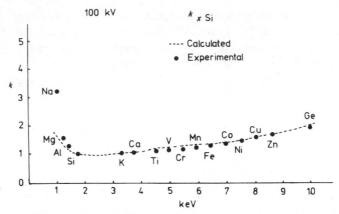

Fig. 3.15. Comparison between experimental and calculated k_{xSi} values as a function of k ($Z < 19$) or k_z ($Z \geq 19$) X-ray energy at 100 kV.[40]

that the average X-ray travels through half the specimen thickness, and is absorbed according to the relationship

$$I = I_0 \exp(\mu \rho t/2 \operatorname{cosec} \theta) \qquad (3.19)$$

When the *ratio* of characteristic X-ray intensities is considered, the *difference* in X-ray absorption coefficients is the important parameter (if two characteristic X-rays have similar values of μ they will be absorbed equally and their intensity ratio will be unaffected). If it is assumed that the average X-ray travels a distance equal to $t/2 \operatorname{cosec} \theta$ where t is the sample thickness and θ the angle between the line of the detector and the sample surface, the observed X-ray intensity ratio I_A/I_B will be modified from that of an infinitely thin specimen I_{0A}/I_{0B} (no absorption)

$$\frac{I_A}{I_B} = \left(\frac{I_{0A}}{I_{0B}}\right) \exp(\mu_{SPEC}^A - \mu_{SPEC}^B)\rho t/2 \operatorname{cosec} \theta \qquad (3.20)$$

where μ_{SPEC}^A and μ_{SPEC}^B are the mass absorption coefficients of the characteristic radiation from elements A and B in a specimen (SPEC) of density ρ. When the absolute value of the exponential in the above equation is equal to $\sim 0 \cdot 1$, absorption in the specimen will alter the observed characteristic X-ray intensity ratio by 10 %; when this is decreased to $\sim 0 \cdot 01$ the characteristic X-ray intensity ratio will be modified by 1 %.

Figure 3.16 shows an application of the technique of analytical electron microscopy to the analysis of carbide particles. The $M_{23}C_6/M_6C$ carbides have been extracted from a medium-carbon steel containing 6 wt % W after

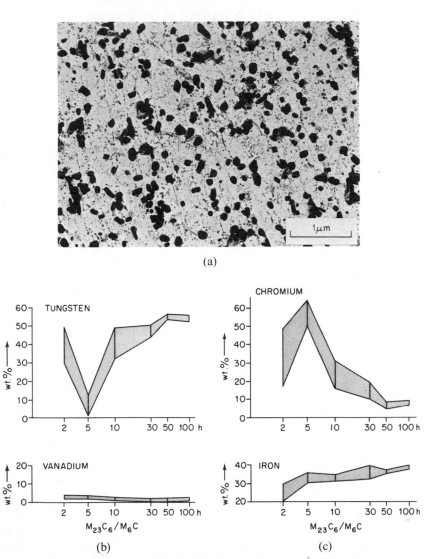

(a)

(b)   (c)

Fig. 3.16.   (a) Electron micrograph of an extraction replica of a medium-carbon steel containing 6 wt % W tempered at 700 °C for 4 h. (b) and (c): Variation in composition of $M_{23}C_6/M_6C$ carbides with tempering time at 700 °C.[41]

various tempering times at 700 °C. Figure 3.16(a) shows a typical carbide distribution and Figs. 3.16(b) and (c) the variation in composition of the precipitates with time at 700 °C.

ACKNOWLEDGEMENT

This chapter was written while the author was a visiting Senior Research Fellow at the Department of Materials Engineering, Monash University, Australia. I would like to express my appreciation for the hospitality which was extended to me during my stay at Monash and for the excellent technical and secretarial assistance which I received when preparing this paper.

REFERENCES

1. P. B. Hirsch, A. Howie, R. B. Nicholson, D. W. Pashley and M. J. Whelan, *Electron Microscopy of Thin Crystals*. Butterworths, London, 1965.
2. N. F. Mott and F. R. N. Nabarro, *Proc. Phys. Soc.*, 1940, **52**, 86.
3. M. F. Ashby and L. M. Brown, *Phil. Mag.*, 1963, **8**, 1083.
4. J. W. Edington, *Practical Electron Microscopy in Materials Science*, Vols. 1–4. Philips Technical Library, The Macmillan Press, London, 1976.
5. P. E. Champness and G. W. Lorimer, *J. Mat. Sci.*, 1973, **8**, 467.
6. R. B. Nicholson and J. Nutting, *Acta Met.*, 1961, **9**, 210.
7. G. W. Lorimer and P. E. Champness, *Phil. Mag.*, 1973, **28**, 1391.
8. K. V. Subba Rao, M.Sc. Thesis, Manchester University, 1971.
9. R. T. DeHoff and F. H. Rhines, Eds., *Quantitative Metallography*. McGraw-Hill, New York, 1968.
10. F. B. Pickering, *The Basis of Quantitative Metallography*. Metals and Metallurgy Trust for the Institute of Metallurgical Technicians, London, 1976.
11. E. Scheil, *Z. Metallk.*, 1935, **27**, 199.
12. J. W. Cahn and R. L. Fulman, *Trans. AIME*, 1956, **197**, 447.
13. R. T. DeHoff, *Trans. Met. Soc. AIME*, 1962, **224**, 474.
14. E. F. Underwood, in *Quantitative Microscopy* (eds. R. T. DeHoff and F. H. Rhines), p. 151. McGraw-Hill, New York, 1968.
15. J. E. Hilliard, *Trans. AIME*, 1962, **224**, 906.
16. G. W. Lorimer, G. Cliff and N. J. Clark, *Developments in Electron Microscopy and Analysis* (ed. J. A. Venables), p. 153. Academic Press, New York, 1975.
17. G. Love, M. G. C. Cox and V. D. Scott, *Developments in Electron Microscopy and Analysis*, Inst. Physics Conf. Ser. No. 36 (ed. D. L. Misell) p. 347, 1977.
18. S. Amelinckx, *The Direct Observation of Dislocations*, p. 193. Academic Press, New York, 1964.
19. P. M. Kelly, A. Jostsons, R. G. Blake and J. G. Napier, *phys. stat. sol.* (a), 1975, **31**, 771.

20. D. Cherns, A. Howie and M. H. Jacobs, *Z. für Natur.*, 1973, **28a,** 565.
21. G. W. Lorimer, N. A. Razik and G. Cliff, *J. Microsc.*, 1973, **99,** 153.
22. P. C. Morris, N. C. Davies and J. A. Treverton, in *Developments in Electron Microscopy and Analysis*, Inst. Physics Conf. Ser. No. 36 (ed. D. L. Misell) p. 377, 1977.
23. M. N. Thompson, P. Doig, J. W. Edington and P. E. J. Flewitt, *Phil. Mag.*, 1977, **35,** 1537.
24. H. L. Fraser, N. J. Zaluzec, J. B. Woodhouse and L. B. Sis, *Proc. 33rd EMSA Meeting*, Las Vegas, 1975 (ed. G. W. Bailey) p. 106.
25. D. J. H. Cockayne, I. L. F. Ray and M. J. Whelan, *Phil. Mag.*, 1969, **20,** 1265.
26. D. J. H. Cockayne, *J. Microsc.*, 1973, **98,** 116.
27. H. Yoshida, D. J. H. Cockayne and M. J. Whelan, *Phil. Mag.*, 1976, **34,** 89.
28. P. E. Champness and G. W. Lorimer, *Phil. Mag.*, 1974, **30,** 357.
29. M. J. Goringe, *J. Microsc.*, 1973, **93,** 95.
30. W. Johnson, *Metals and Materials*, Feb., 1975, 21.
31. G. W. Lorimer and P. E. Champness, in *Proc. Third International HVEM Conference*, 1974 (eds. P. R. Swann, C. J. Humphreys and M. J. Goringe) p. 301. Academic Press, New York.
32. E. P. Butler and P. R. Swann, in *Physical Aspects of Electron Microscopy and Microbeam Analysis* (eds. B. Siegel and D. R. Beaman) p. 129. John Wiley and Sons, London, 1975.
33. P. R. Swann, in *Electron Microscopy and the Structure of Materials* (ed. G. Thomas) p. 898. University of California Press, Berkeley, 1972.
34. M. J. Makin and J. V. Sharp, *J. Mat. Sci.*, 1968, **3,** 360.
35. W. Kossel and G. Möllenstedt, *Ann. Phys.*, 1939, **5,** 113.
36. J. A. Eades, in *Developments in Electron Microscopy and Analysis*, Inst. Physics Conf. Ser. No. 36 (ed. D. L. Misell) p. 283, 1977.
37. K. E. Cooke, *ibid.*, p. 431.
38. J. J. Goldstein, J. L. Costley, G. W. Lorimer and S. J. B. Reed, *Proc. Workshop on Analytical Electron Microscopy*, SEM 1977, **1,** IIT Research Institute, Chicago, p 315.
39. G. Cliff and G. W. Lorimer, *J. Microsc.*, 1975, **103,** 203.
40. G. W. Lorimer, S. A. Al-Salman and G. Cliff, in *Developments in Electron Microscopy and Analysis*, Inst. Physics Conf. Ser. No. 36 (ed. D. L. Misell) p. 369, 1977.
41. R. J. Tunney, G. W. Lorimer and N. Ridley, *Met. Sci.*, 1978, **12,** 271.

4

The Application of Transmission Electron Microscopy to the Study of Plastic Deformation

M. H. LORETTO and R. E. SMALLMAN

The University of Birmingham, UK

4.1 INTRODUCTION

The aim of this chapter is to review the role played by transmission electron microscopy in studying the plastic deformation of crystals. To appreciate this role it is essential that readers have a clear understanding of the mode whereby electron-optical images and diffraction patterns are obtained from crystals, and of the theory that enables the images of defects to be related to the actual geometry of the crystal defects. The theories that have been developed to interpret the images are outlined in Section 4.3 and it is clear from these theories that the images can only be meaningfully interpreted if the experimental conditions under which images are obtained are precisely defined.

This chapter deals in the main with the bright-field technique rather than dark-field microscopy. It should be emphasised, however, that the basic theory for defect analysis is common to both techniques and that the division is entirely arbitrary. In addition, it should be pointed out that this division is somewhat unfortunate since it has led to the misconception that there are fundamental differences between the two techniques, as, for example, in the determination of the Burgers vector of a dislocation, whereas in fact precisely the same theory is used. Bright-field electron microscopy is carried out by inserting an aperture in, or close to, the back focal plane of the objective lens so that (ideally) only those electrons travelling in the direction of the incident beam† are allowed to contribute to

---

† Typically the aperture is 50 μm in diameter and inelastically scattered electrons within this region also make up the image.

the final image. In dark-field microscopy the incident beam is tilted so that only one diffracted beam is allowed to form the image. Thus, all beams but one are excluded and the crystal planes are, therefore, deliberately not resolved; any contrast in the image arises solely from variations in intensity in the transmitted beam at the bottom surface of the crystal. These variations in intensity in the direct beam will be caused by any factor that locally alters the intensity in any diffracted ray and it thus follows that the bright-field image is dependent on (among other things) the orientation of the crystal, i.e. on the diffraction conditions. The success of transmission electron microscopy (TEM) is due, to a great extent, to the fact that it is possible to define these diffraction conditions by obtaining a diffraction pattern from the same small volume of crystal (as small as 1 μm diameter) as that from which the electron micrograph is taken. As a result, it is possible to obtain the crystallographic and diffraction information necessary to interpret electron micrographs. To obtain a selected area diffraction pattern (SAD) an aperture is inserted in the plane of the first image so that only that part of the specimen which is imaged within the aperture can contribute to the diffraction pattern.† The power of the diffraction lens is then reduced so that the back focal plane of the objective is imaged, and then the diffraction pattern, which is focussed in this plane, can be seen. Clearly, the objective aperture must be removed so that the diffracted beams are allowed to contribute to the diffraction pattern.

   In view of the fact that the image obtained from a crystal is determined by the diffraction conditions (which are usually adjusted so that only one set of planes is diffracting strongly) it is appropriate to deal first with the information obtainable from a diffraction pattern, to explain more fully why this information is needed, and how the diffraction conditions are controlled experimentally. We will then outline the basis of the simple two-beam (i.e. the one strong diffracted beam and the directly transmitted beam) kinematical equations and the more complex two-beam dynamical equations, which describe the intensity of the two beams and the way they are used to interpret electron micrographs. It will then be shown how the information derived from the electron diffraction patterns can be introduced into these equations, and how changes in the various experimental parameters influence the image contrast. In the next two sections some applications of these basic ideas to the study of the nature of defects that occur in plastically deformed crystals will be considered, and

---

† Errors can arise[1] in selected area diffraction due to focussing error and spherical aberration of the objective.

some of the work that has led to an increased understanding of plastic deformation will be described. Finally, some recent developments, mainly in the field of high-voltage electron microscopy (HVEM), will be discussed.

4.2 DIFFRACTION PATTERNS

Transmission electron diffraction patterns contain (at least) three parameters which are essential in the interpretation of the associated electron micrographs. The first of these we will denote as **B**, the electron beam direction, and we will follow the convention that the electron beam direction is considered as upwards from the specimen so that it is acute to the upward drawn foil normal, **N**. The determination of **B** is important because the electron micrograph obtained is a projection of a three-dimensional object where the projection direction is **B**. Every direction on a micrograph, such as the crystallographic direction parallel to a dislocation line, is only a projected direction and this projected direction will in general vary as **B** is changed. In order to determine the true crystallographic direction of any line, we need, therefore, to obtain several micrographs in different electron beam directions, and then, using standard crystallographic techniques, determine the true direction.

The second and third of the parameters obtained from a diffraction pattern are the indices **g** of the crystal plane (hkl), which is set near to, or at, the Bragg angle, and the precise deviation from the Bragg angle of this crystal plane. This deviation parameter is usually represented by **s**. All three of these parameters are conveniently represented on the well-known Ewald sphere construction (if a plane, (hkl) is at the Bragg condition g_{hkl} lies on the sphere). Thus, Fig. 4.1 shows **B**, **g** and **s** in terms of this construction. The solution of a diffraction pattern for **B** is given in Chapter 3.

This solution for **B** (which assumes that reflections that are excited correspond to planes at the Bragg angle) is only approximate, since reflections can be excited when the reciprocal lattice point lies significantly off the Ewald sphere.[1] A more accurate solution can be obtained from Kikuchi lines which are present in the diffraction patterns obtained from somewhat thicker specimens. It is convenient to discuss the origin and geometry of Kikuchi lines at this stage, since the determination and experimental selection of **g** and **s** are both done using Kikuchi lines.

A typical electron diffraction pattern showing Kikuchi lines is given in Fig. 4.2, from which it can be seen that these consist of a pair of bright and dark lines associated with each reflection, spaced [g] apart. A simple theory

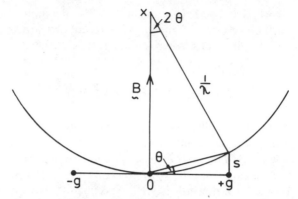

Fig. 4.1.   Ewald sphere construction illustrating the relation between the electron beam direction **B**, the diffracting vector **g**, and the deviation parameter **s**.

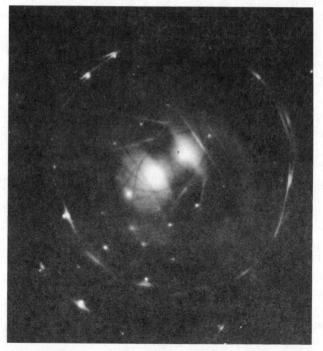

Fig. 4.2.   Electron diffraction pattern obtained from a crystal of silicon which was sufficiently thick and perfect to give Kikuchi lines.

for the formation of these lines was put forward by Kikuchi in 1928,[2] and although this theory is somewhat inadequate in many respects it is nevertheless a useful way of describing the origin of these lines. The explanation given by Kikuchi assumes that at some point in the crystal (P on Fig. 4.3) an inelastic scattering event takes place so that point P acts as a spherical source of electrons. It is further assumed that there is a negligible

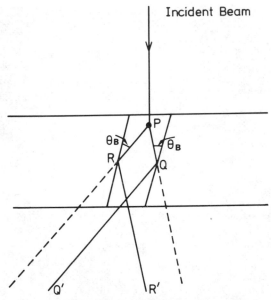

Fig. 4.3.   The mechanism for the formation of Kikuchi lines.

change in wavelength associated with this inelastic scattering event so that the Bragg angle for these electrons is identical with the Bragg angle of the incident electrons. Thus, for the set of planes in Fig. 4.3, those electrons travelling in the directions $\overline{PQ}$ and $\overline{PR}$ will be Bragg-diffracted at Q and R and give rise to rays in the directions QQ′ and RR′. Since the electrons in the beam RR′ originate from the scattered ray PR, this beam will be less intense than QQ′, which contains electrons scattered through a smaller angle at P. Because P is a spherical source this rediffraction at points such as Q and R gives rise to cones of rays, which, when they intersect the plate and viewing screen, approximate to straight lines, as seen in Fig. 4.2. There is of course a pair of lines (2θ apart) for each set of Bragg planes and several pairs of lines are visible on Fig. 4.2. Because of the way in which the Kikuchi lines are formed they behave as if rigidly attached to the crystal; as the crystal is tilted the Kikuchi lines will move. It is this fact that is important in the accurate determination of **B**, the selection of **g** and the measurement of **s**.

An accurate determination of **B** can be carried out from Kikuchi lines as shown in the following example (Fig. 4.4). When the electron beam is nearly along [hkl], as in Fig. 4.4, it is clear that the intersections of the bisectors of

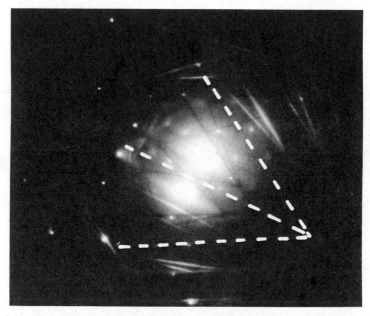

Fig. 4.4.   The method used to determine the electron beam direction from Kikuchi lines. For explanation see text.

(non-parallel) pairs of Kikuchi lines meet at a point (as shown by dashed lines in Fig. 4.4) and that this point is at a measurable distance from the direct beam. This distance can be expressed in degrees, since the Bragg angles are known, and we thus have a direct measure of the deviation of the electron beam direction from the crystallographic direction defined by the point of intersection of these dashed lines. This direction is given by the cross products of g_1 and g_2 where g_1 and g_2 are the indices of the planes giving rise to the pairs of Kikuchi lines. Figure 4.2 is an example in which the electron beam direction is very close to a low index direction and lines bisecting the Kikuchi bands would meet close to the direct beam, as would be expected from the above argument.

As indicated earlier, the value of **s** can be obtained from a diffraction pattern that contains Kikuchi lines. Thus, from the Ewald sphere construction (cf. Fig. 4.1) the value of **s** at symmetry (i.e. when $-g_{hkl}$ and $+g_{hkl}$ are both equally outside the sphere and the planes (hkl) are therefore vertical) is given by

$$\mathbf{s} = -\mathbf{g}^2\lambda/2$$

Therefore, for the diffraction pattern shown in Fig. 4.5, the value of **s** is given by simple proportion, since at symmetry the first-order Kikuchi lines bisect the distance between the direct beam and the diffracted beam. In practice, since the first-order Kikuchi line is usually diffuse it is preferable to use the sharper, second- or third-order lines, which are marked on Fig. 4.5, on which to make the actual measurements. The expression above can be modified if this second-order line is used, or alternatively, the first-order line can be marked in since it is separated by θ_B from the second-order line. The importance of the value of **s** in contrast is discussed in Section 4.3.

The selection of the diffracting conditions used to image the crystal defects can be controlled using Kikuchi lines. Thus the planes (hkl) are at the Bragg angle when the corresponding pair of Kikuchi lines passes through 000 and $\mathbf{g}_{hkl}$, i.e. **s** = 0, as is evident from the above equation. Tilting of the specimen so that this condition is maintained (which can be done quite simply, using modern double-tilt specimen stages) enables the operator to select a specimen orientation with a close approximation to two-beam

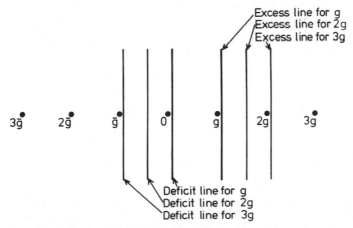

Fig. 4.5. A schematic diffraction pattern from which the value of **s** may be determined, as described in the text.

conditions† which, as is discussed in Section 4.3, enables micrographs to be taken under well-defined conditions and hence to be more simply interpreted. Similarly, various electron beam directions can be selected using Kikuchi lines as a navigational aid. The way in which this is done and the way in which successive diffracting vectors are selected in a systematic manner are best illustrated by reference to Kikuchi maps; these will therefore be introduced at this stage.

4.2.1 Kikuchi Maps

When a crystal is oriented so that the electron beam is parallel to a high-symmetry direction, then, since all crystal planes can give rise to pairs of

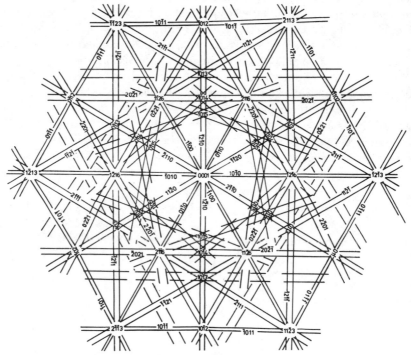

Fig. 4.6. A schematic Kikuchi map for an HCP crystal.

† Typically this would correspond to an electron beam direction 5° away from a low index pole. For example, tilting a face-centred cubic (FCC) crystal $6\frac{1}{2}°$ away from $\mathbf{B} = [011]$ whilst maintaining $\mathbf{g} = 200$ at the Bragg angle would mean that the new electron beam direction would be [045] or [054]. No other reflections smaller than the 0,10,8 type would be excited in an FCC crystal.

Kikuchi lines, this symmetry will be easily recognizable in the Kikuchi pattern. This feature is illustrated in Fig. 4.2 and in the schematic Kikuchi map shown in Fig. 4.6 where (low index) Kikuchi lines are given for a hexagonal close-packed (HCP) crystal encompassing electron beam directions around [0001].

The indexing of the Kikuchi lines and corresponding reflections is shown more clearly in Fig. 4.7 for the case of the electron beam close to [011]. Figure 4.7(a) shows an indexed Kikuchi map and Fig. 4.7(b) a diffraction pattern indexed from the map. Note that the labelling of electron beam directions is the opposite to that on stereograms, e.g. the [013] electron beam direction is indicated on Fig. 4.7 where [031] would appear on a corresponding stereogram. A simple crystal model can be used to show that this interchange of the indexing between Kikuchi maps and stereograms is required in order to index correctly the various electron beam directions for

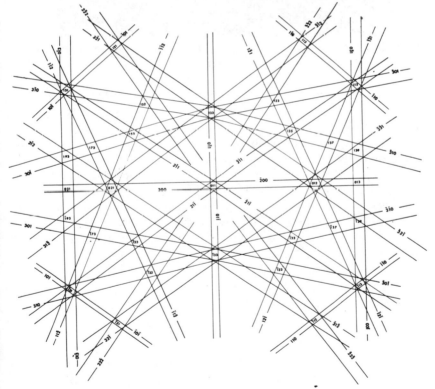

Fig. 4.7(a). An indexed Kikuchi map for a BCC crystal, centred on [011].

Fig. 4.7(b).   A diffraction pattern from a BCC crystal indexed from the Kikuchi
map of Fig. 4.7(a).

successive diffraction patterns. The main value of Kikuchi maps lies in the
fact that unless the *sense* of rotations between successive electron beam
directions is recorded, information is lost. For example, the electron
diffraction patterns obtained by rotating a crystal from [001] about [1$\bar{1}$0]
will result in the electron beam becoming parallel to either [113] or [$\bar{1}\bar{1}$3],
depending on the sense of the rotation. The way in which the Kikuchi
pattern changes during tilting removes this ambiguity and thus allows
internally consistent indexing of electron beam directions and diffracting
vectors, which is essential if defect analysis is to be carried out correctly.
Kikuchi maps are also invaluable in studying less symmetrical crystals
simply as a labour- and time-saving device. Once an experimental Kikuchi
map has been obtained by building up a composite map,[3] then all
diffraction patterns can be solved by inspection.

4.3   IMAGE CONTRAST

In the previous section we have seen how information is obtained from a
diffraction pattern. From the present standpoint the main aim of obtaining
this information is to use it to interpret the associated electron micrograph

by applying the theories of image contrast[1] that have been developed. An electron micrograph is simply a highly magnified image of the variation in intensity of the direct beam (BF) or diffracted beam (DF) across the bottom surface of the foil. The problem of interpreting contrast then becomes one of calculating these intensities. Two simplifying assumptions are usually made in these calculations: firstly, that the intensity on the bottom surface can be obtained by dividing the crystal into columns (~ 20 Å in diameter) and calculating the intensity on the bottom surface of each, independently of each other; secondly, it is generally assumed that only two beams need be considered, the direct beam and a diffracted beam. It should be clear from the preceding section that this two-beam approximation can be closely approached experimentally, at least at 100 kV. As will be seen later, both of these approximations are considered adequate at 1000 kV.

4.3.1 Kinematic Theory of Image Contrast

The kinematic theory is the simplest theory that has been developed to predict image contrast. Apart from the assumptions mentioned earlier, several further assumptions are made in this theory, the most important of which is that the intensity scattered into the diffracted beam is considered to be small so that any rediffraction of the (assumed) weak diffracted beam is therefore neglected. The method used to calculate the amplitude, and hence the intensity, in each column of a perfect crystal is to consider a column whose axis is the diffracted beam. The column is imagined to be divided into slabs perpendicular to the direction of the diffracted beam and the amplitude at the bottom of the column from each of these slabs calculated by assuming that the latter act as Fresnel zones. The total amplitude is then obtained by summing these contributions over the thickness, t, of the crystal. The expression for ϕ_g, the amplitude of the diffracted beam, is then found, for an incident amplitude $\phi_0 = 1$, to be

$$\phi_g = (\Pi i/\xi_g) \int_0^t \exp(-2\Pi i s z \, dz)$$

where ξ_g (termed the extinction distance for the operative reflection, $\mathbf{g}$) is

$$\xi_g = \Pi v_c \cos \theta / \lambda F(\theta)$$

and v_c is the volume of a unit cell, λ is the wavelength of electrons, $F(\theta)$ is the structure factor for the reflection, and $\mathbf{s}$ is the deviation parameter. If $\mathbf{s}$ is not a function of z then

$$\phi_g = (\Pi i/\xi_g)(\sin \Pi t s/\Pi s) \exp[-(i\Pi t s)]$$

and
$$|\phi_g|^2 = I_g = (\Pi^2/\xi_g^2)(\sin^2 \Pi ts/\Pi^2 s^2)$$

At $s = 0$, this expression reduces to $(\Pi t/\xi_g)^2$ and thus for $t > \xi_g/\Pi$ the diffracted intensity I_g is predicted to exceed the incident intensity. Despite this obvious limitation, the theory has the virtue of simplicity and, provided its approximate nature is appreciated, it is a worthwhile way of considering contrast. In particular, the weak-beam technique[4] (which is discussed later) deliberately uses very large values of s and in such work the kinematic theory has been found to give reasonable agreement with experiment and with the more sophisticated theory.

In order to calculate the influence of imperfections (dislocations, stacking-faults, etc.) on the contrast, the displacement **R** of a unit cell from its true lattice position needs to be calculated. The expression for the amplitude is easily shown to be

$$\phi_g = (\Pi i/\xi_g) \int_0^t \exp\{-2\Pi i(sz + \mathbf{g}.\mathbf{R})\}\,dz$$

The major difficulty in solving this equation lies in the expression for **R**. If isotropic elasticity theory is used then a considerable simplification results. Thus, if we consider, for example, the simple case of a screw dislocation at a depth y below the surface of a crystal, as shown in Fig. 4.8, the displacement at the point p in the column indicated is given by[5]

$$R = (b/2\Pi)\{\tan^{-1}(z - y)/x\}$$

and thus

$$\phi_g = (\Pi i/\xi_g) \int_0^t \exp\{-2\Pi i(sz + (g.b/2\Pi)\tan^{-1}[(z - y)/x]\}\,dz$$

Solving this equation (by computer) for columns either side of a dislocation enables a profile of intensity to be obtained. By computing profiles for different assumed values of y through the thickness of the foil it is possible to build up a picture of the intensity associated with a dislocation through the foil.[1] Because of the simplifying assumptions made, the theory cannot be used to predict details of image contrast but the principle behind the 'invisibility criteria' that are used for the determination of the Burgers vector of a dislocation is clear from a comparison of the respective expressions for ϕ_g for the perfect and imperfect crystal; the only difference is the term $\mathbf{g}.\mathbf{R}$. Hence, if **R** is perpendicular to **g** then $\mathbf{g}.\mathbf{R} = 0$ and the dislocation is predicted to be invisible. Since, in isotropic elasticity theory, the only displacements associated with a screw dislocation are in the

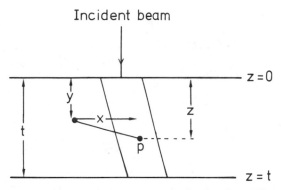

Fig. 4.8. The parameters used to calculate the amplitude of the diffracted beam on the bottom surface of a crystal. The column considered is at a distance x from a screw dislocation, which is at a depth y below the surface of the crystal, which is of thickness t.

direction of **b**, this invisibility criterion is simply saying that all those planes that contain **b** are not tilted, and hence diffract electrons as if the dislocation were not present. Because the displacements are more complex around an edge dislocation, then the conditions for invisibility are more restrictive and it is easy to show that **g**.**b** and **g**.**b**∧**u** must both be zero for invisibility where **u** is a vector along the dislocation line. For a mixed dislocation, isotropic elasticity theory predicts that some contrast is always expected. Before discussing this aspect it is appropriate to deal with the dynamical theory of diffraction contrast since residual contrast is too detailed a concept to discuss either in terms of the kinematic theory or in terms of isotropic elasticity.

4.3.2 Dynamical Theory of Image Contrast

The two-beam dynamical theory of image contrast[1] as developed by Howie and Whelan can be derived in a very similar way to that used for the kinematic theory, using a wave-optical formulation. In this theory account is taken of the rediffraction of the direct and diffracted beams and hence the expressions for ϕ_0 and ϕ_g are coupled. The equations derived for the variation with depth of diffracted and transmitted amplitude are then

$$\mathrm{d}\phi_0/\mathrm{d}z = (\Pi i\phi_0/\xi_0) + (\Pi i\phi_g/\xi_g)\exp(2\Pi isz)$$

$$\mathrm{d}\phi_g/\mathrm{d}z = (\Pi i\phi_g/\xi_g) + (\Pi i\phi_0/\xi_g)\exp(-2\Pi isz)$$

where all the symbols have the meanings used earlier and ξ_0, like ξ_g, has the

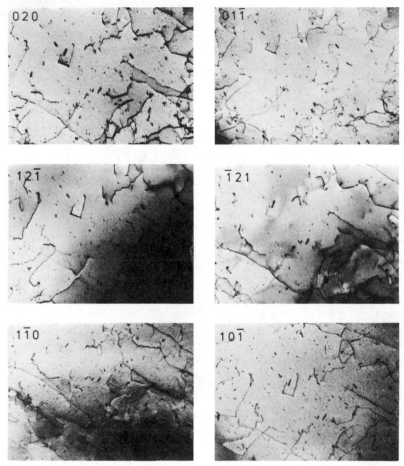

Fig. 4.9. Bright-field electron micrographs showing the effect of varying **g** on the contrast associated with dislocations. The diffracting vectors are indicated.

dimensions of length, is proportional to $f(0)$, the atomic scattering amplitude for zero angle, and is therefore related to the refractive index of the crystal. By comparing experimental electron-optical images of crystals with images computed on this theory it has been shown that it is necessary to replace the terms involving extinction distances by complex quantities in order to obtain agreement between observation and prediction, so that $1/\xi_g$ is replaced by $(1/\xi_g) + (i/\xi_g')$ and $1/\xi_0$ is replaced by $(1/\xi_0) + (i/\xi_0')$. This is really an empirical way of recognising that absorption occurs, i.e. that some electrons (other than those in the diffracted beam) do not contribute to the image since they are scattered outside the aperture when it is around the direct beam. The formal justification of this substitution is beyond the scope of this chapter. In precisely the same way as in the kinematic theory the contrast from imperfect crystals introduces an extra phase factor $2\Pi\mathbf{g}\cdot\mathbf{R}$ where $\mathbf{R}$ is the displacement term.

A particularly useful form of the two-beam dynamical equations for an imperfect crystal is

$$d\phi_0'/dz = \Pi i\phi_g'\{(1/\xi_g) + (i/\xi_g')\}$$

$$d\phi_g'/dz = \Pi i\phi_g\{(1/\xi_g) + (i/\xi_g')\} + 2\Pi i\phi_g'\{s + g(d\mathbf{R}/dz)\}$$

Here the primes are used to indicate that different phase factors are used in the expressions for the amplitudes. Since we are interested only in intensities this does not matter and it is customary to drop the primes.

The above coupled differential equations make it very clear that the displacement term $g(d\mathbf{R}/dz)$ (where $d\mathbf{R}/dz$ is equivalent to a local tilting of the reflecting plane) locally changes the effective value of s and thus the contrast observed will depend both on s and on $g(d\mathbf{R}/dz)$; clearly, when $g(d\mathbf{R}/dz)$ is zero the defect will be invisible and when s is large then large values of $g(d\mathbf{R}/dz)$ will be needed to give significant contrast. The micrographs shown in Fig. 4.9 show the effect of varying $\mathbf{g}$. Although this chapter is concerned in the main with bright-field, the more interesting and important differences in the visibility of dislocations arise at large values of s in dark-field; this aspect will be discussed in relation to the weak-beam technique in the following section.

4.4　TECHNIQUES OF DEFECT ANALYSIS

Perhaps the most important aspect of the application of transmission electron microscopy to the study of plastic deformation is the fact that extremely precise information can be obtained concerning the defects that are imaged in micrographs. In this section some of the more important techniques that enable this type of information to be obtained will be discussed.

4.4.1 Determination of the Burgers Vector of Dislocations
As already indicated, the application of isotropic elasticity theory to the visibility of dislocations leads to the conclusion that if $\mathbf{g}\cdot\mathbf{b} = 0$ a screw

dislocation will be invisible, and in addition if $\mathbf{g} \cdot \mathbf{b} \wedge \mathbf{u}$ also equals zero an edge dislocation will be invisible; mixed dislocations are predicted to show some contrast under all diffracting conditions. These criteria for invisibility were widely used for many years, and are indeed still used, but it is important to realise their limitations and where doubt arises more sophisticated techniques should be applied. Problems may arise in determining Burgers vectors when a material is significantly elastically anisotropic and also when weak images are associated with diffracting conditions when $\mathbf{g} \cdot \mathbf{b} \neq 0$. Weak images are in fact obtained when large values of $\mathbf{s}$ are used to image dislocations (as would be expected intuitively from the discussion in Section 4.3) since the term $\mathbf{g} \cdot d\mathbf{R}/dz$ is additive to $\mathbf{s}$ and this has led to erroneous identification of the Burgers vector of dislocations,[6] by assuming that the near-invisibility condition produced corresponded to $\mathbf{g} \cdot \mathbf{b} = 0$. This difficulty arises when diffracting vectors are used to image dislocations for which a variety of values of $[\mathbf{g} \cdot \mathbf{b}]$ are possible for the family of dislocations imaged. For example, if a 321 diffracting vector is used to image dislocations in a BCC metal then for the dislocations with $\mathbf{b} = \frac{1}{2}\langle 111 \rangle$, values of $[\mathbf{g} \cdot \mathbf{b}]$ of: 3, 2, 1 and 0 are possible. The images which correspond to conditions of $\mathbf{g} \cdot \mathbf{b} = 3$ and 2 are broad and intense at small values of $\mathbf{s}$ and the images corresponding to $\mathbf{g} \cdot \mathbf{b} = 1$ which are relatively weak, may be confused with the condition $\mathbf{g} \cdot \mathbf{b} = 0$. In practice a large value of $\mathbf{s}$ is selected experimentally so that the broad intense images corresponding to $\mathbf{g} \cdot \mathbf{b} = 2$ and 3 become narrower and more acceptable. The consequence of this is that images for which $[\mathbf{g} \cdot \mathbf{b}] = 0$ and 1 become indistinguishable from each other. With the large tilts available there seems no necessity to use large diffracting vectors for determining the Burgers vector of dislocations, but if such vectors are used then it is essential to ensure that several different types of $\mathbf{g}$ are used and also a number of different values of $\mathbf{s}$. An additional, extremely valuable check on whether a weak image is associated with the condition $\mathbf{g} \cdot \mathbf{b} = 0$ is simply to reverse $\mathbf{g}$. Since (as is discussed later) the image[1] of a dislocation moves from one side of the actual position of the dislocation to the other as $(\mathbf{g} \cdot \mathbf{b}) \mathbf{s}$ changes sign; if $|\mathbf{g} \cdot \mathbf{b}| = 0$ the image will not move.

The only reliable method of interpreting contrast from dislocations, in cases of doubt, is to compute images for the experimental conditions used and compare them with the appropriate experimental images for assumed Burgers vectors. This technique, termed image matching, has been developed recently by Head and co-workers and a book devoted to image matching has been published[7] to which the reader is referred for details. The programs used take into account the elastic anisotropy of crystals and a typical example showing the degree of agreement between experiment and theory is illustrated in Fig. 4.10. It should be mentioned that the information required for computing images—e.g. g, B, s, N (the foil normal), u (the true direction of the line defect)—are best obtained in the manner outlined in the earlier sections, using Kikuchi maps as a navigational aid.

4.4.2 Determination of Dislocation Density

The determination of dislocation density is an important aspect of deformation studies and TEM has been used extensively to obtain this information. It is clear that the diffracting conditions need to be controlled for such measurements since some dislocations may be out of contrast. The method usually used is to measure the number of intersections that the images of dislocation lines make with a series of random lines drawn on a micrograph. To obtain the dislocation density from such a measurement requires a knowledge of the foil thickness (Section 4.4.3). Alternatively, the number of dislocation intersections with the foil surfaces can be counted and hence the number of dislocations threading per unit area obtained. To convert this figure to a dislocation density per unit volume requires some knowledge concerning the preferred direction of dislocations with respect to the foil surface. It has been shown[8] that the best conversion factor in the absence of a detailed knowledge of this anisotropy is that the density per unit area should be doubled to obtain the density per unit volume.

4.4.3 Stereo-microscopy and Thickness Determinations

One of the disadvantages in dealing with electron micrographs of complex dislocation structures arises from the overlap of images resulting from the fact that the micrograph is a projection of a three-dimensional object. An extremely valuable technique for separating such overlap is that of stereo-microscopy since it enables a three-dimensional image to be viewed. Stereo-micrographs are simply taken by using the same values of g and s for micrographs taken in two different electron beam directions separated by about $15°$ (although the precise angle is not critical but it does depend upon foil thickness). It should be evident that the taking of stereo pairs is another application of Kikuchi maps. If the pair of stereo-micrographs is viewed in a stereo-viewer then the details of dislocation interactions, for example, are revealed far more clearly than on the individual micrographs.

If the two surfaces are decorated with islands of gold then the foil thickness (in the 'mean' direction of the electron beam) can be obtained with

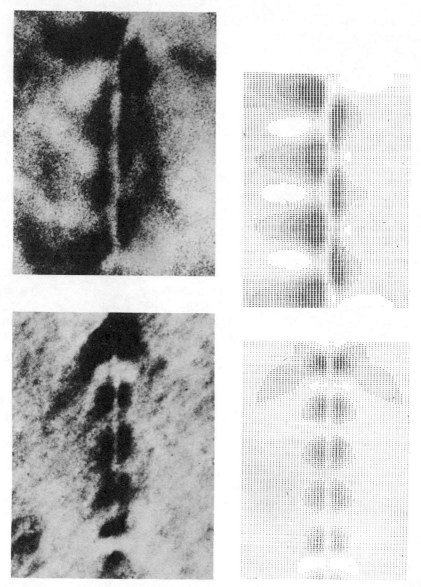

Fig. 4.10.   Experimental (left) and computed (right) images for a screw dislocation
β-brass. The computed images show the contrast predicted when $\mathbf{g} \cdot \mathbf{b} = 0$ and
these match the experimental images. (After Head *et al.*, see Ref. 7.)

a calibrated viewer and thus if the foil is tilted equally either side of the foil normal the true foil thickness can be measured directly. When the foil normal is not known accurately but the foil thickness is required, then stereo-micrographs are again useful because the ends of dislocations that lie in the same foil surface can be joined and from the directions of two such lines $\mathbf{u}_1$ and $\mathbf{u}_2$, the foil normal obtained as $\mathbf{u}_1 \wedge \mathbf{u}_2$. The ends of dislocations that lie on opposite surfaces can also be joined (on other micrographs taken in different electron beam directions) and hence the true directions of the lines obtained. Thus the foil thickness can be obtained from the relation:

$$t = L_\mathrm{m} \left[\frac{(h_1 h_3 + k_1 k_3 + l_1 l_3)}{(h_3^2 + k_3^2 + l_3^2)^{1/2}} \right]$$

$$\times \left\{ h_1^2 + k_1^2 + l_1^2 - \frac{(h_1 h_2 + k_1 k_2 + l_1 l_2)^2}{(h_2^2 + k_2^2 + l_2^2)} \right\}^{-1/2}$$

where $[h_1 k_1 l_1]$ is the true direction of a line in foil, $[h_2 k_2 l_2]$ is $\mathbf{B}$, the electron beam direction $[h_3 k_3 l_3]$ is $\mathbf{N}$, the foil normal, and L_m is the measured length of line running along $[h_1 k_1 l_1]$.

4.4.4 The Determination of the Nature of Dislocation Loops

A large number of dislocation loops and dipoles are formed during the plastic deformation of metals (and also by quenching or irradiation) and hence it is of interest to determine the nature of these defects, i.e. whether they are vacancy or interstitial in nature. To appreciate the method used to differentiate between these interstitial and vacancy loops it is necessary to define the direction of the Burgers vectors of such loops. If we use the FS/RH convention and take the positive sense of the loop as clockwise when viewed from above a loop, then the Burgers vector of an interstitial loop has a component upwards and a vacancy loop downwards with respect to the plane of the loop. This is illustrated in Fig. 4.11. The size of the image of a loop is predicted to be greater when $(\mathbf{g} . \mathbf{b})s$ is positive and thus the image is predicted to change size as the sign of $(\mathbf{g} . \mathbf{b})s$ is changed. Thus for a pure-edge loop the direction of $\mathbf{b}$ can be obtained by (a) determining the direction ($+$ or $-$) of $\mathbf{b}$ using the invisibility criteria discussed earlier, (b) determining the plane of the loop by carrying out large tilting experiments, and (c) reversing the sense of $\mathbf{g}$ (preferably with several different diffracting vectors) and noting the sense of image shift.

Complications arise when the loops are not pure edge[9] but the experimental technique is basically the same. It should be noted that since

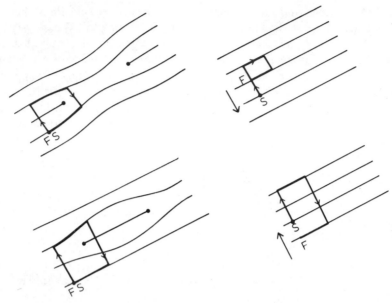

Fig. 4.11.   The FS/RH convention used in defining **b**. (See text.)

the magnitude of the image shift is related to the size of (**g . b**) this is one application where large diffracting vectors should be used.

An example showing the change in loop size on reversing **g** is given in Fig. 4.12.

4.4.5 The Determination of the Nature of Stacking-faults
Stacking-faults appear on micrographs as a series of dark and bright fringes parallel to the line of intersection of the foil surface and the plane of the

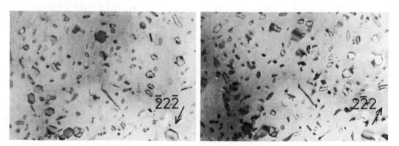

Fig. 4.12.   Images of dislocation loops in ion-implanted silicon showing the influence of the sign of **g** on the size of the images of the loops.

fault. The form of the Howie–Whelan equations used for calculating fault contrast is different from that given earlier, since a stacking-fault in a foil simply results in the bottom part of the crystal being displaced with respect to the top half. A suitable form for the equations is

$$d\phi_0/dz = \Pi i(1/\xi_0 + i/\xi_0')\phi_0 + \Pi i(1/\xi_g + i/\xi_g')\phi_g \exp(2\Pi isz + 2\Pi ig.\mathbf{R})$$
$$d\phi_g/dz = \Pi i(1/\xi_0 + i/\xi_0')\phi_g + \Pi i(1/\xi_g + i/\xi_g')\phi_0 \exp(-2\Pi isz - 2\Pi ig.\mathbf{R})$$

Calculations have shown[1] that the intensity of the outermost fringe is determined by the parameter $\alpha = 2\Pi\mathbf{g}.\mathbf{R}$. (With very thin foils this rule breaks down.)[10] Thus, if the fault in an FCC metal is considered to be formed by removing (intrinsic) or adding (extrinsic) a layer of material and if $\mathbf{R}$ is defined as the sense of movement of the bottom of the crystal necessary to accommodate this fault then images computed on the two-beam dynamical theory show that if $\mathbf{g}.\mathbf{R} = +\frac{1}{3}$ the outermost fringe is white, and if $\mathbf{g}.\mathbf{R} = -\frac{1}{3}$ the outermost fringe is black. As in the case of a loop, it is necessary to determine the upward drawn normal of the plane on which a fault lies, determine the value of $\mathbf{g}.\mathbf{R}$ for a variety of reflections and hence determine the nature of the fault. An example of this analysis is shown in Fig. 4.13 where the faults A, B and C were found by trace analysis† to lie on planes with upward drawn normals [111], [1$\bar{1}$1] and [$\bar{1}$11]. The diffracting vector for Fig. 4.13 is $\bar{1}\bar{1}$1 and the values of $\mathbf{g}.\mathbf{R}$ for intrinsic and extrinsic faults are given in Table 4.1. All these faults are deduced to be intrinsic, since the observed contrast is consistent with that expected for intrinsic faults.

TABLE 4.1
Analysis of Stacking Faults in a Face-centred Cubic Crystal

| Fault | Values of $\mathbf{g}.\mathbf{R}$ | | Value of $\mathbf{g}.\mathbf{R}$ deduced from outermost fringe |
	Intrinsic	Extrinsic	
A	$-\frac{1}{3}$	$+\frac{1}{3}$	black, therefore $-\frac{1}{3}$
B	$+\frac{1}{3}$	$-\frac{1}{3}$	white, therefore $+\frac{1}{3}$
C	$+\frac{1}{3}$	$-\frac{1}{3}$	white, therefore $+\frac{1}{3}$

† The upward drawn normal of the fault planes can also be determined by imaging the fault in DF as well as in BF. The fringe which corresponds to intersection with the top surface of the foil appears similar in the two micrographs (for the same **g**) and this is sufficient information to remove any ambiguity in indexing the {111} fault plane.

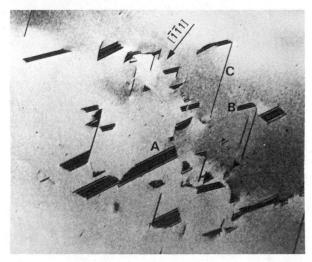

Fig. 4.13.   Electron micrograph of stacking-faults in a Au–Sn alloy. The intensity of the outermost fringe defines whether $\mathbf{g} \cdot \mathbf{R} = +\frac{1}{3}$ or $-\frac{1}{3}$ and thus whether the faults are intrinsic or extrinsic. For full explanation see text.

It should be noted that thin precipitates will give displacement fringe contrast which can be very difficult to distinguish from stacking-fault contrast. The fact that stacking-faults are expected to be invisible when $\mathbf{g} \cdot \mathbf{R} =$ an integer, i.e. the component of $\mathbf{R}$ along $\mathbf{g}$ is an integral number of inter-planar spacings, appears to offer a possible way of distinguishing faults and precipitates, since for a precipitate it is unlikely (but not impossible) that $\mathbf{g} \cdot \mathbf{R}$ will be an integer. However, recent work has shown that the displacement associated with a stacking-fault is not simply $\frac{1}{3} \langle 111 \rangle$; there is, in addition, a small supplementary displacement due to the local change in stacking, which gives rise to contrast even when $\mathbf{g} \cdot \frac{1}{3} \langle 111 \rangle$ is integral.[52]

4.4.6 Weak-beam Technique

One of the limiting factors in the analysis of dislocations is the fact that images of dislocations are usually $\gtrsim 100$ Å wide; this clearly limits the detail that can be observed. It has been shown[4,11] that if $\mathbf{s}$ is set at a value exceeding about 2×10^{-2} Å^{-1} very narrow images of about 15–20 Å are obtained, which, in dark field, are very much above background intensity. The principle behind the technique is apparent from the form of the contrast equations for an imperfect crystal where the displacement term is additive to s, as pointed out earlier. Thus, if a very large value of s is selected when imaging a defect, only those displacements that are very large can give rise to a strong diffracted beam, i.e. can rotate the planes associated with the diffracting vector too close to the Bragg angle. Images at large s will therefore be narrower than at small s because at large s only those planes very close to the core will be bent sufficiently to cause significant contrast. The volume of crystal giving rise to this enhanced diffraction is small and can be detected most easily in dark field since the background intensity[1] then approaches zero at large values of s; thus in dark field the signal-to-background intensity is optimised. In view of the low absolute intensity of weak-beam images, exposures of up to 30 sec are commonplace and this necessitates a drift-free stage if high resolution is to be realised. Some weak-beam images are difficult to see on the viewing screen and this can lead to focussing difficulties. However, all focussing can, in fact, be carried out on the corresponding bright-field image, if the microscope is equipped with dark-field deflector coils, since the following procedure can be carried out: first (a) deflect the diffracted beam g, which is to be used to image the crystal, down the optic axis of the microscope; then (b) return to BF and set this particular diffracting vector near to the Bragg angle, i.e. so that the Kikuchi pairs corresponding to this reflection are set as in Fig. 4.2; (c) return to DF with g down the optic axis, when 3g will be close to the Bragg angle† and hence the value of s for g is very large; (d) return to BF and select the area of interest, adjusting the focus and crystal tilt as appropriate; and finally, (e) return to DF and take the weak-beam micrograph.

If the weak-beam micrograph is too strong, i.e. if the images of dislocations are too broad, then the value of s can be increased in BF, the image refocussed and stage (e) repeated. An example of the decrease in image width associated with weak-beam microscopy is shown in Fig. 4.14.

This technique, known as the g/3g technique, requires that the specimen be tilted if the sense of g is to be reversed—as is required in loop analysis (cf. Section 4.4.4). If the crystal is set at the symmetry condition in BF (cf. Fig. 4.1) and g now deflected down the optic axis, 2g will be at the Bragg condition and a g/2g weak-beam condition will be satisfied. If $\bar{g}$ is deflected down the optic axis then $\bar{g}/2\bar{g}$ is satisfied. Thus a $\pm g$ (weak-beam) condition can be achieved without tilting the specimen, which is useful when imaging dipoles or small loops, since any change in image size can only be due to the sense of image shift and not to specimen tilt.

† This is evident from Fig. 4.5 since Kikuchi lines will not move (as long as the crystal is not tilted) and when g is deflected the direct beam and 3g are displaced so that the appropriate deficit and excess lines pass close to 0 and 3g respectively.

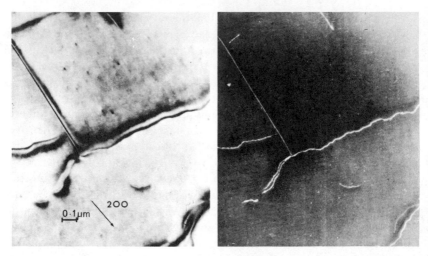

Fig. 4.14.   Bright-field (left) and weak-beam (right) micrographs of dislocations of $\mathbf{b} = \langle 111 \rangle$ in NiAl, illustrating the decrease in image width with weak-beam images. The diffracting vector is indicated and the value of $\mathbf{s}$ for the weak-beam image is $2 \cdot 4 \times 10^{-2} \, \text{Å}^{-1}$.

4.5   APPLICATION TO PLASTIC DEFORMATION

This section will deal with some selected areas in the study of the plastic deformation of crystals in which advances have been made by applying the theory and techniques outlined in the previous sections.

4.5.1   Determination of Stacking-fault Energy

One of the most important parameters which influence the work-hardening and general plastic properties of metals is the stacking-fault energy. Many methods for determining γ, the stacking-fault energy, have been suggested, but those that use TEM are without question the most easily interpreted. All other methods developed so far are not capable of quantitative analysis and are useful only on a comparative basis. For example, the method based on the analysis of textures[12] requires normalising to values obtained by direct observation.

The method due initially to Whelan[13] based on measurements of the radius of curvature of extended three-fold nodes is illustrated in Fig. 4.15. The radius, R, indicated can be approximated by the expression used by Whelan, $\gamma = T/R$, where T is the line tension of the dislocation, usually taken to be the line energy. Brown[14] has developed an expression that enables a much more accurate value of γ to be obtained from the measurement of the node width w as indicated in Fig. 4.15. The determination of γ in fact requires a knowledge of the character of the dislocations making up the node, i.e. the angle between **b** and **u**. There is clearly a disadvantage in using a complex configuration such as a node to determine γ, since, as Brown's work shows, the relation between w and γ

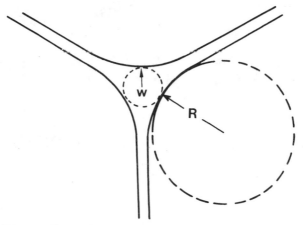

Fig. 4.15.   Diagram illustrating the parameters w, the width, and R, the radius, of a three-fold node.

requires a computer even on the assumption that the material is elastically isotropic. However, it must be recognised that there are two very distinct advantages. Firstly, there is the obvious advantage that the splitting of the partial dislocation is larger at a node than for the same partials away from the node—this splitting factor is about four and thus extends the range of direct observation of such splitting. Secondly, if a node is symmetrical, i.e. the character of each dislocation entering the node is the same, then the angle between the dislocations making up the node should be 120° if the node is in equilibrium. Any node that is not in equilibrium is recognisably unacceptable for a measurement of γ since some factor, such as internal stress or solute impedance, must be influencing the geometry and hence the separation of the partials; the resulting value of γ is also liable to be incorrect. If the measurements of γ are made on isolated dislocations then there is no similar way (when conventional BF imaging is used) of recognising whether or not the separation of the partial dislocations is a

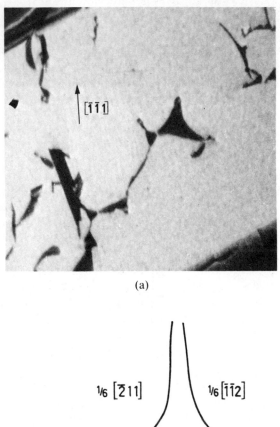

(a)

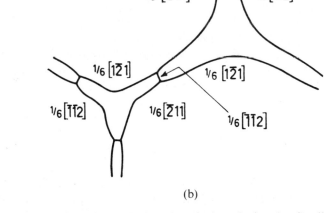

(b)

Fig. 4.16.   (a) Extended intrinsic and extrinsic nodes in a Au–Sn alloy (after M. H. Loretto, *Phil. Mag.*, 1964, **10**, 467). (b) Schematic of (a).

non-equilibrium separation. However, as discussed later, weak-beam imaging removes this problem.

Networks of alternately extended and contracted nodes are commonly observed and generally it is the intrinsic nodes that are extended, and for which Brown's calculations apply. Occasionally,[15] extended intrinsic and extrinsic nodes are as shown in Fig. 4.16(a) and, schematically, in Fig. 4.16(b) but the shape and the variation in size, as well as the more complex geometry of the extrinsic nodes, make it impossible to obtain an estimate of γ_e. However, a defect closely related to that shown in Fig. 4.16 is the fault pair.[16] The geometry of these defects is relatively simple and it has been shown that the value of γ_i obtained applying anisotropic elasticity theory to these fault pairs is very similar to that using isotropic theory to extended nodes. In addition it has been established from measurements on these defects that $\gamma_e = 1 \cdot 1 \gamma_i$.

The usefulness of these measurements on extended nodes and fault pairs is clearly limited to values of γ for which a separation is observable and until the recent application of weak-beam microscopy to the determination of γ this meant that the minimum size of node measurable as extended was about $R = 300$ Å. Thus, of the pure FCC metals, only the stacking-fault energy of silver could be obtained directly and the determination of γ for other pure metals was done by extrapolation of node measurements on alloys. Recent work using weak-beam microscopy has shown that values of γ can be obtained from direct observations of nodes in copper and gold; an example of the advantage of weak-beam microscopy applied to the observation of extended nodes is shown in Fig. 4.17. Figure 4.17(a) shows the bright-field image and Fig. 4.17(b) the weak-beam image. The separation of the partial dislocations in these higher stacking-fault energy materials is only about 50 Å and since the approximation of linear elasticity is used to obtain γ, the limit of the technique appears to have been reached.[17-20] Thus the application of weak-beam imaging of isolated dislocations has revealed that the relation between the degree of dissociation and the character of a dislocation (i.e. the deviation from screw orientation) deviates from that predicted from linear anisotropic elasticity when the dissociation is small, i.e. ~4 nm. In this weak-beam work the improved resolution enables jogs and constrictions to be observed, hence the cause of some of the experimental scatter in node size and dissociation of isolated dislocations can be identified. If measurements of dissociation are made sufficiently far away from constrictions, the scatter in γ obtained from isolated dislocations is found to be less than from extended nodes.[20] With this improvement it seems that measurements of separation as a function of character of isolated dislocations is the most reliable technique for determining γ.

There are other methods of determining γ using TEM which can be applied to higher stacking-fault-energy materials. Perhaps the most powerful of these is the method based on the rate of annealing of dislocation loops.[21] The rate of annealing of faulted dislocation loops is far greater than for unfaulted loops and the difference in rates can be used to determine

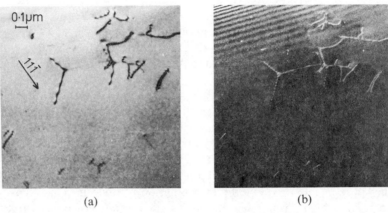

(a) (b)

Fig. 4.17. (a) A bright-field and (b) a weak-beam micrograph of an extended node in a stainless steel (Bampton *et al.*, *Acta Met.*, 1978, **26**, 39).

γ_i from single-faulted loops and γ_e from double-faulted loops.[22] The method has been used for aluminium for which $\gamma_i = 135 \, \text{mJ/m}^2$ but cannot be used for low γ materials such as gold and silver because jog nucleation and propagation are the important parameters in determining shrinkage rate.[23,24] This method is therefore complementary to the extended node observations. In addition, it has been shown that during plastic deformation of metals, stacking-fault tetrahedra and triangular Frank loops can be formed by a dislocation reaction rather than by point defect nucleation, and values of γ_i can be obtained from a measurement of the size of the smallest Frank loop and the largest tetrahedron.[25,26] This method is limited to those metals that form tetrahedra during deformation, and these have been observed in copper, silver and gold, from which values of γ_i have been obtained in approximate agreement with weak-beam observations.

Some other methods of determining γ have been used, such as measurements[20,27,28] on faulted dipoles, but weak-beam measurements on isolated dislocations as a function of character and loop annealing methods are probably the best available.

4.5.2 Determination of the Mechanism of Kinking in Single Crystals

The phenomenon of kinking in single crystals has been studied for many years since Orowan's initial work[29] on cadmium. It is known that suitable single crystals kink when compressed in a direction such that the primary slip systems have a resolved shear stress close to zero acting on them. For example, zinc and cadmium tend to kink when compressed in directions nearly perpendicular to [0001] and NiAl tends to kink when compressed along [001]. In the case of zinc, macroscopic observations have shown that crystals aligned precisely perpendicularly to [0001] do not kink at room temperature but kinking occurs at 77 K in such crystals.[30] On the other hand kinking does not occur at 77 K in NiAl but may occur at temperatures between about 300 K and 1000 K, depending on the strain rate. This complex interaction between temperature, strain rate and orientation has been interpreted[31-33] by correlating macroscopic observations with a detailed study using TEM of the nature and type of dislocations present in specimens that had deformed either by kinking or by some other mechanism. Thus, it has been shown in the case of NiAl that whenever a crystal does deform by kinking the only mobile dislocations are those with $\mathbf{b} = \langle 100 \rangle$. At low temperatures—77–300 K—the dislocations responsible for the plastic strain have $\mathbf{b} = \langle 111 \rangle$ and it has been shown,[34] using weak-beam microscopy, that these dislocations exist as a pair of dislocations of $\mathbf{b} = \frac{1}{2}\langle 111 \rangle$ separated by about 50 Å of antiphase boundary. The predominance of the operation of $\langle 111 \rangle$ slip at 77 K can then be understood since the temperature dependence of the flow stress of dislocations of $\mathbf{b} = \frac{1}{2}\langle 111 \rangle$ will be smaller than that for dislocations of $\mathbf{b} = \langle 100 \rangle$. At higher temperatures ($\gtrsim 300$ K) the crystals either kink or deform uniformly without any observable surface markings but, as shown by TEM, the Burgers vector of all the mobile dislocations that are observed after kinking or uniform deformation is $\mathbf{b} = \langle 100 \rangle$. This somewhat surprising result, taken together with the strain rate dependence of kinking, showed that glide of dislocations of $\mathbf{b} = \langle 100 \rangle$ was the dominant mechanism of deformation when kinking was observed and that kinking was due to geometrical softening, but that climb of dislocations of $\mathbf{b} = \langle 100 \rangle$ was responsible for the shape change associated with uniform deformation. Similarly, it has been shown that the only dislocations present in zinc crystals that kink are those with $\mathbf{b} = \frac{1}{3}\langle 11\bar{2}0 \rangle$ and that kinking is due to geometrical softening of the $\langle 11\bar{2}0 \rangle (0001)$ slip systems. Crystals that do

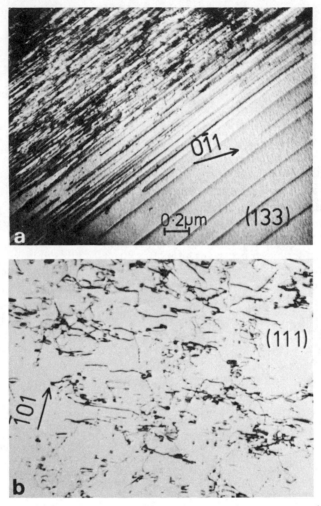

Fig. 4.18.   Electron micrographs of the dislocation microstructures observed in NiAl deformed along $\langle 001 \rangle$ at (a) 77 K, (b) 1050 K.

not kink deform by the movement of $\frac{1}{3}\langle 11\bar{2}3 \rangle$ dislocations. Since the temperature dependence of the flow stress of $\frac{1}{3}\langle 11\bar{2}0 \rangle$ dislocations is very small, the increased probability of kinking observed at low temperatures can also be explained. Examples of the different dislocation microstructures observed in NiAl as the temperature of straining is increased from 77 K up to 1050 K are shown in Fig. 4.18. The micrograph obtained from a specimen deformed at 77 K is a good illustration of the different mobility of edge and screw components of dislocations at low temperature; the mobile edge component has clearly dragged out long screw segments.

4.5.3 Deformation Substructures

Electron microscope observations show that during strain hardening a substructure is often formed by the motion and interaction of dislocations. Thus, after extensive plastic deformation, the majority of dislocations are arranged in complex cell walls, separated by regions of very low dislocation density, with the cell walls tending to be both on the slip planes and on planes normal to the slip planes. In BCC materials, cell walls on {110}, {112} and {123} have been observed, and in copper on {110}, {100} and {111}. There is conflicting evidence concerning the nature of the lattice rotations across the dislocation networks formed by plastic deformation, but the detailed investigation by Steeds[35] shows that the alternate misorientations of the tilt type arose from the accumulation of forest dislocations of opposite signs, and that the twist boundaries arose from the interaction of the primary and coplanar slip systems. The mechanism of formation of tangles and cell walls during deformation is still not clearly understood, but, in general, tangle formation appears to result from the generation of secondary dislocations and their interaction with the primary piled-up groups. This area of deformation behaviour will undoubtedly be furthered by *in situ* dynamic observations (see Section 4.6).

In low stacking-fault energy metals the dislocations are observed to remain in planar groups due to the difficulty of cross-slip and, as a consequence, no regular cell walls are formed. The presence of second-phase particles also affects the distribution of dislocations and an important parameter appears to be the particle size. No dense tangles of secondary dislocations are produced in the vicinity of small particles—less than a few hundred Ångströms in diameter—whereas localised tangling occurs in the vicinity of large particles.

In high stacking-fault energy materials, the dislocation distribution changes with increasing plastic strain from a fairly regular arrangement at small strains to a well-defined cell structure after 5–10 % elongation. In both FCC and BCC materials it is observed that cell size decreases progressively with increasing strain until a minimum value is reached. When the cell size becomes constant, deformation proceeds by the motion of dislocations between cell walls with no effective tangling of dislocations inside the cells. Dislocations may still continue to accumulate at the cell walls and raise the flow stress, either by increasing the internal stress asssociated with the wall or by reducing the effective source length of any dislocation which can be unpinned from the cell wall.

Stacking-fault energy has been established to be an important parameter in determining the creep rate, $\dot{\varepsilon}$, of FCC metals and alloys, which can be expressed as

$$\dot{\varepsilon} = \left(\frac{\sigma}{G}\right)^n \gamma^m D$$

where σ is the applied stress, G is the shear modulus, γ is the stacking fault energy and D is the diffusivity. The stacking-fault exponent m is usually about 3 and the stress exponent n about 5 or higher, depending on the dislocation density of the matrix controlling the creep rate by climb. In dispersion-hardened alloys the strengthening role of particles has been attributed to direct particle/dislocation interactions and also to the effect of particles in stabilising the dislocation substructure by inhibiting recrystallisation.

In nickel-based alloys containing ThO_2 where the creep rate is controlled by the climb of the dislocation substructure, Wilcox and Clauer[36] have observed a stress exponent of 7 (similar to that of Ni-alloys not containing particles) and a dependence of the creep rate on γ. However, in Ni-alloys that do not possess an inherent substructure, particle-bypassing controls the creep behaviour[37] and this is achieved at all temperatures of deformation by the cross-slip of screw dislocations to establish a double jog configuration at the particle, which then climb. Thus climb is rate-controlling at all temperatures but is affected by composition-dependent pipe diffusion at low temperatures ($< 0.5\ T_m$) and by bulk lattice diffusion at high temperatures.

In fatigue deformation the substructures[38] produced are also affected by stacking-fault energy: low stacking-fault energy materials contain planar arrays of dislocations and dipoles whereas high stacking-fault energy materials produce either dipole arrays or cell structures somewhat more regular than those developed in tension. In high stacking-fault energy metals the dislocation substructures develop at low plastic strain amplitudes, i.e. those that will give a fatigue life of 10^6 cycles differ from those developed at higher plastic strain amplitudes. The dislocation arrangement consists predominantly of dense mats of edge dislocations which accumulate on the primary slip planes and in walls perpendicular to the primary slip planes. The dislocation debris lies along the traces of the critical conjugate slip planes and in copper contains numerous Lomer–Cottrell dislocations and Frank dipoles. At higher strain amplitudes, the character of the substructure is changed from a dipole structure to a three-dimensional cell structure. The cell size is determined by the strain amplitude and temperature; the cell walls are relatively simple dislocation configurations, but at lower temperatures the walls contain a high density of loops and dipoles.

4.5.4 The Deformation of Alloys Containing Particles

The understanding of the factors that control the strength of alloys containing particles is an important field of plastic deformation in which TEM has played a central role. It is apparent that a fundamental understanding of particle-strengthened materials is likely only if the way in which dislocations interact with the particles can be studied. This can be done only with TEM.

In 1948 Orowan suggested that if an alloy contains non-deformable particles the shear stress would be determined by the dislocations bowing around the particles in their glide plane, leaving Orowan loops encircling each particle. He deduced that the yield stress, τ, alloy should be given by $\tau = \tau_m + (2T/bD_s)$ where τ_m is the matrix yield stress, T the line tension of the dislocation, b the Burgers vector of the dislocation and D_s the mean planar spacing of the particles. This expression has been modified to take account of complicating factors such as the variation of line tension with dislocation character, and reasonable agreement has been obtained[39] between measurements of yield stress and the modified Orowan stress in copper–silica alloys. However, it has also been suggested[40] that dislocations may bypass particles by cross-slip and the influence of this on the yield stress needs to be considered.

In addition to increasing the yield stress, non-shearable particles will also influence the work-hardening rate. The rate of work-hardening at low strains is perhaps consistent with the formation of Orowan loops as discussed by Fisher et al.[41] Recent results obtained using TEM have been used to account for the yield stress and to formulate new theories of work-hardening; some of these results will be discussed here.

The extensive research work at Oxford[42] and Cambridge[43] has established the dislocation structure in a series of alloys containing Al_2O_3 or SiO_2 particles of various sizes. In particular, Humphreys has deformed single crystals of $Cu–SiO_2$, $Cu–Al_2O_3$, and taken micrographs of slip-plane sections that illustrate the complexity of the structure and how it depends on the stacking-fault energy. A characteristic feature, which Humphreys finds at low strains, is the presence of rows of prismatic loops parallel to the primary Burgers vector when the stacking-fault energy is high. At lower

Fig. 4.19.   HVEM micrograph of specimen of aged Cu–Cr–SiO$_2$ deformed to a shear strain of 0·10. Note the complex dislocation array around the large SiO$_2$ particle. **g** is indicated and the electron beam direction is near [1$\bar{2}$1].

values of stacking-fault energy (and associated solution-hardening) Orowan loops are formed. As pointed out by Hirsch and Humphreys, this observation is consistent with the idea that the prismatic loops are produced by cross-slip rather than prismatic punching. Indeed a detailed analysis of the type of loop, i.e. interstitial or vacancy, and the arrangement and their sizes has been carried out and the results are consistent with calculations for a cross-slip mechanism.[44]

Work by Brown and Stobbs[43] has shown that around larger SiO$_2$ particles ($>3000\,\text{Å}$ diameter) more complex dislocation arrays are produced which tend to inhibit the formation of prismatic loops, and they have examined various possible stress relaxation mechanisms around particles.

Long[45] has examined the defect structure in deformed specimens of aged Cu–Cr–SiO$_2$ which contained large (mean size $\sim$500 nm) particles of SiO$_2$. The Cr precipitates control the flow stress and allow the defect structure around these large particles to be retained in thin foils. An example showing the complex structure around some large particles is given in Fig. 4.19, and detailed analysis has revealed that the structure is in very close agreement with that predicted by Brown and Stobbs.

Although there is still some uncertainty about the mechanism by which some of these structures are developed during deformation, the direct observation of the dislocations associated with non-shearable particles has provided a firm basis for theories. The direct observation of voids at particle interfaces[46] is a further important contribution to the understanding of the plastic properties of these alloys. Thus, in dispersion-hardened materials, and metals containing second-phase particles as impurity, it is found that the particle often acts as a site for nucleating voids that grow with increasing strain, eventually leading to ductile failure. The voids are nucleated as a result of the back stress on the particle/matrix interface created by the secondary dislocations generated near the particle to accommodate the shear strain in the matrix. Thus, the degree of bonding between the particle and the matrix is extremely important in governing the ductility of a material. This degree of bonding between particle and matrix may be determined[47] by TEM from observations of voids on particles annealed to produce an equilibrium configuration by measuring the contact angle θ of the matrix surface to the particle surface.

A large amount of work has also been carried out on the defect structure in alloys that contain ordered particles, and in age-hardening alloys that have in common the feature that the particles can be sheared.

4.6 FUTURE DEVELOPMENTS

In the field of deformation studies there is at least one aspect of electron microscopy where exciting developments are expected and indeed are already being realised. This is in the field of high-voltage electron microscopy (HVEM).

The installation of high-voltage microscopes in the UK has resulted in a large increase in activity in this field. One of the obvious advantages of HVEM is, of course, the fact that thicker specimens can be examined. This has, for example, made it possible to study dislocation–particle interactions at the large particle sizes[43,45] studied by Brown and Stobbs. It has also led to a better appreciation of the three-dimensional arrangement of dislocations, and, generally, the greater thickness has enabled bulk behaviour to be more easily appreciated. In addition to these direct applications of HVEM, a considerable amount of work has been carried out using dynamic straining experiments to further the knowledge of the behaviour of dislocations during straining. There are many practical difficulties involved in these experiments. It is tempting to use the thickest foils possible in an attempt to limit the role of the surfaces and thus use the highest accelerating voltage available, but observations show how rapidly climb will occur when metals are examined at voltages significantly above the threshold for damage. Even if low beam intensities are used at high voltages, so that climb is not easily observable, it would seem to limit the confidence that can be placed on the directly observed dislocation behaviour if the experimental conditions are such that climb can occur as a result of point defect production by the electron beam. It is thus essential to limit the voltage to below the threshold for damage in general studies using HVEM. In addition, it is clearly desirable to have a tensile stage which is capable of extremely smooth operation, and in which the strain rate can be reduced to zero so that the crystal can be examined under stress when required. A stage has been built[48] which is a modified version of the double-tilting straining stage designed by Vesely,[53] and incorporates piezo-electric crystals to provide totally vibration-free straining, and the ability to stop and start straining at will. Experiments using this stage have been carried out on molybdenum single crystals[48-50] (with and without carbides), Cu/Cr, Cu/SiO$_2$[51] and stainless steel.[52]

In the experiments on Mo it has been shown that although the foil normal, N, is an important factor in determining which slip systems operate, as shown by Vesely,[53] it is possible, by varying N, to isolate this factor. Thus the slow-moving screw dislocations control the strain rate if they cannot escape at the foil surface. If the foil surface contains b then the screw dislocations can escape and the fast-moving edge dislocations dominate the deformation. For example, if a crystal is stressed along [010] and the foil normal is [101] then those dislocations with $b = \frac{1}{2}[\bar{1}11]$ and $\frac{1}{2}[\bar{1}\bar{1}1]$ will dominate the deformation despite the fact that the $\frac{1}{2}[111]$ and $\frac{1}{2}[1\bar{1}1]$ dislocations have the same applied tensile stress acting on them. If the tensile axis is maintained as [010] and the foil normal changed to [100] all four $\frac{1}{2}\langle 111 \rangle$ dislocations are found to operate to a similar extent.

Detailed recordings can be made on video-tape or 35-mm film using time-lapse photography of the many dislocation interactions that are observed,

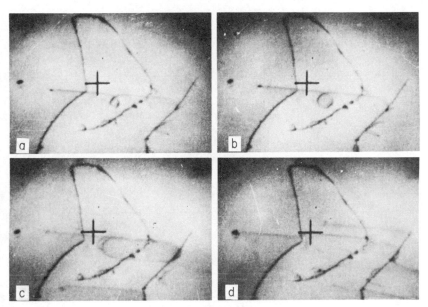

Fig. 4.20. A series of micrographs taken under load during the deformation of a Mo single crystal. For explanation see text. Accelerating voltage 800 kV.

and an example of the generation of screw dislocations from a cusp in a dislocation is shown in Fig. 4.20(a)–(d). The importance of molybdenum carbide particles in the generation of dislocations in molybdenum that has been heat treated in various ways, has been clarified by observing the early stages of plasticity in the HVEM (cf. Fig. 4.21). In strain-aged molybdenum the only sources of dislocation were associated with the carbide particles, and if there was only sufficient carbon to pin the grown-in dislocation then the single crystals are inherently brittle. Decarburised molybdenum was seen to be ductile; the few grown-in dislocations were able to generate dislocations since they were not pinned. On the other hand the ductility of carburised specimens was seen to be due to the high density of unpinned dislocations associated with the large carbide particles. The ability to observe directly the early stages of plasticity in these specimens enabled a straightforward explanation of the corresponding macroscopic obser-vations but *in situ* observations are not always useful in throwing light on bulk behaviour. For example, specimens of Cu–SiO$_2$ have been deformed *in situ* and the interaction of dislocations and SiO$_2$ particles observed. In most cases the dislocations simply cross-slip to avoid the SiO$_2$ and the

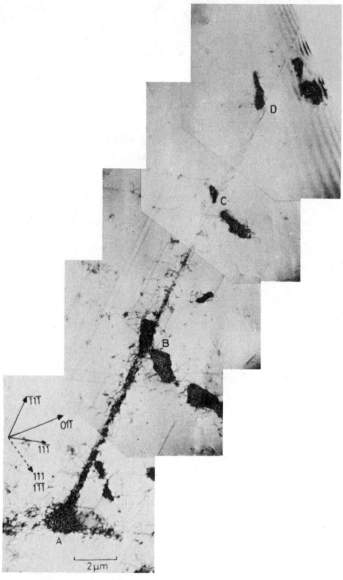

Fig. 4.21.   Montage taken from a specimen of Mo containing carbide particles which are seen to generate dislocations at the particle/metal interface.

structure observed after straining is not typical of specimens deformed in bulk. In fact a large number of dipoles are formed and it appears that the dislocations do not always totally avoid the particles by cross-slip; a jog is formed by cross-slip and the dipole is then dragged out. Although similar dipoles are observed in specimens deformed in the bulk, they are observed in far greater number in specimens deformed *in situ*. No Orowan loops were observed and it seems that this is one material where *in situ* straining has not been very useful. On the other hand, *in situ* straining of aged Cu–Cr developed structures that were identical to those observed in bulk-deformed specimens. Clearly the spacing of the particles in an aged alloy is such that for most of the line length of dislocation the influence of the stress-free foil surfaces is irrelevant.

The formation of ε- and α-martensite during the *in situ* deformation of stainless steel is also being studied and the observations compared with the formation of martensite during cooling. The fact that straining (or cooling) can be arrested at any stage during the formation of martensite enables detailed analysis to be carried out on those defects that are formed before the subsequent observation of HCP and BCC diffraction maxima. In this way it is hoped to understand the nucleation mechanism of martensite in these steels.

There is no doubt that *in situ* straining is a very difficult experimental technique but the work briefly summarised above indicates the potential of this approach.

REFERENCES

1. P. B. Hirsch, A. Howie, R. B. Nicholson, D. W. Pashley and M. J. Whelan, *Electron Microscopy of Thin Crystals*. Butterworths, London, 1965.
2. S. Kikuchi, *Jap. J. Phys.*, 1928, **5**, 83.
3. G. Thomas, *Modern Diffraction and Imageing Techniques in Materials Science* (ed. S. Amelinckx). North-Holland, Amsterdam, 1970.
4. D. J. H. Cockayne, I. L. F. Ray and M. J. Whelan, *Phil. Mag.*, 1969, **20**, 1265.
5. W. T. Read, *Dislocations in Crystals*. McGraw-Hill, New York, 1953.
6. L. K. France and M. H. Loretto, *Proc. Roy. Soc. A*, 1968, **307**, 83.
7. A. K. Head, P. Humble, L. M. Clarebrough, A. J. Morton and C. T. Forwood, *Computed Electron Micrographs and Defect Identification*. North-Holland, Amsterdam, 1973.
8. G. Shoeck, *J.A.P.*, 1962, **32**, 1745.
9. D. M. Maher and B. L. Eyre, *Phil. Mag.*, 1971, **23**, 409.
10. A. K. Head, *Aust. J. Phys.*, 1969, **22**, 569.
11. D. J. H. Cockayne, *J. Microsc.*, 1973, **98**, 116.

12. I. L. Dillamore and R. E. Smallman, *Phil. Mag.*, 1965, **12**, 191.
13. M. J. Whelan, *Proc. Roy. Soc. A*, 1968, **249**, 114.
14. L. M. Brown, *Phil. Mag.*, 1964, **10**, 441.
15. M. H. Loretto, *Phil. Mag.*, 1964, **10**, 467.
16. P. C. J. Gallagher, *phys. stat. solidi*, 1966, **16**, 95.
17. R. C. Perrin and E. J. Savino, *J. Microsc.*, 1973, **98**, 214.
18. W. M. Stobbs and C. H. Sworn, *Phil. Mag.*, 1971, **24**, 1365.
19. A. Gomez, D. J. H. Cockayne, P. B. Hirsch and V. Vitek, *Phil. Mag.*, 1975, **31**, 105.
20. C. Bampton, I. P. Jones and M. H. Loretto, *Acta Met.*, 1978, **26**, 39.
21. P. S. Dobson and R. E. Smallman, *Proc. Roy. Soc. A*, 1966, **293**, 423.
22. P. S. Dobson, P. J. Goodhew and R. E. Smallman, *Phil. Mag.*, 1967, **16**, 9.
23. L. M. Clarebrough, P. Humble and M. H. Loretto, *Can. J. Phys.*, 1967, **45**, 1135.
24. I. A. Johnston, P. S. Dobson and R. E. Smallman, *Crystal Lattice Defects*, 1969, **1**, 47.
25. M. H. Loretto, L. M. Clarebrough and R. L. Segall, *Phil. Mag.*, 1965, **11**, 459.
26. T. Jossang and J. P. Hirth, *Phil. Mag.*, 1966, **13**, 657.
27. J. W. Steeds, *Phil. Mag.*, 1967, **16**, 785.
28. A. J. Morton and C. T. Forwood, *Crystal Lattice Defects*, 1973, **4**, 165.
29. E. Orowan, *Nature*, London, 1942, **149**, 643.
30. J. J. Gilman, *J. Metals*, 1954, **200**, 621.
31. H. L. Fraser, M. H. Loretto and R. E. Smallman, *Phil. Mag.*, 1973, **28**, 667.
32. H. L. Fraser, R. E. Smallman and M. H. Loretto, *Phil. Mag.*, 1973, **28**, 651.
33. E. G. Tapetado and M. H. Loretto, *Phil. Mag.*, 1974, **30**, 515.
34. R. G. Campany, M. H. Loretto and R. E. Smallman, *J. Microsc.* 1973, **98**, 174.
35. J. W. Steeds, *Proc. Roy. Soc. A*, 1966, **292**, 343.
36. B. A. Wilcox and A. H. Clauer, *Metal Sci. J.*, 1969, **3**, 26.
37. J. Hancock, I. L. Dillamore and R. E. Smallman, *Metal Sci. J.*, 1972, **6**, 153.
38. J. D. Embury, *Strengthening Methods in Crystals* (eds. A. Kelly and R. B. Nicholson) p. 331. Elsevier Mat. Sci. Series, Elsevier, Amsterdam, 1971.
39. R. Ebeling and M. F. Ashby, *Phil. Mag.*, 1966, **13**, 805.
40. P. B. Hirsch, *J. Inst. Metals.*, 1957, **86**, 7.
41. J. C. Fisher, E. W. Hart and R. H. Pry, *Acta Met.*, 1953, **1**, 336.
42. P. B. Hirsch and F. J. Humphreys, *Proc. Roy. Soc. A*, 1970, **318**, 45, 73.
43. L. M. Brown and W. M. Stobbs, *Phil. Mag.*, 1971, **23**, 1185, 1201.
44. M. F. Duesbery and P. B. Hirsch, *Conf. Fundamental Aspects of Dislocation Theory.*, Nat. Bureau of Standards, Washington, 1969.
45. N. J. Long, Ph.D Thesis, University of Birmingham, 1977.
46. I. G. Palmer and G. C. Smith, *Second Bolton Landing Conf., Oxide, and Dispersion Strengthening*, Washington. Gordon and Breach, London, p. 253, 1963.
47. J. Hancock, I. L. Dillamore and R. E. Smallman, *Sixth Plansee Conference*, p. 476, 1969.
48. R. G. Campany, M. H. Loretto and R. E. Smallman, *Metal Science*, 1976, **10**, 253.
49. R. G. Campany, R. E. Smallman and M. H. Loretto, *Metal Science*, 1976, **10**, 261.

50. M. G. Peach, A. Kumar, M. H. Loretto and R. E. Smallman, *HVEM Conference*, Kyoto, Japan, pp. 411–14, 1977.
51. E. Haque, Ph.D. Thesis, University of Birmingham, 1977.
52. J. W. Brooks, M. H. Loretto, and R. E. Smallman, *Ninth International Conf. Electron Microscopy*, Toronto, pp. 624–5, 1978.
53. D. Vesely, *Phys. Stat Solidi*, 1968, **29**, 675.

5

Scanning Electron Microscopy

C. G. VAN ESSEN

Patscentre International, Royston, UK

5.1 INTRODUCTION

In 1938 M. von Ardenne constructed an instrument in which an electron beam passed through a thin specimen and exposed a photographic plate. It was quite different from the transmission electron microscope, however, in that the beam was focussed to as fine a point as possible at the specimen, and scanned the surface point by point. The transmitted beam also scanned the plate point by point, but the plate was moved so that the distances scanned were much greater, thus producing a magnified image. This instrument was the first scanning electron microscope (SEM) and it is interesting that the examination of thin specimens, which was largely neglected during the years of SEM development, has been revived and has become an important technique.

The modern development of the SEM has been almost entirely due to work started in Cambridge by C. W. Oatley in 1948. With the staff of the Engineering Department, and a succession of research students, the instrument was improved to a stage where a commercial version seemed viable. Cambridge Scientific Instruments sold the first 'Stereoscan' in October 1964 to the DuPont company. Fourteen years later, Cambridge offered three different 'Stereoscan' models. At least a dozen other manufacturers are now in business.

The reason why the SEM proved so popular was originally the strikingly three-dimensional manner in which the surfaces of solid specimens could be imaged. But many other modes of operation are possible which provide information (on a very small scale if required) about useful properties of the specimen such as chemical elements, electric potential, magnetic fields and crystallography. Further applications are the subject of vigorous research. The revival of interest in thin specimens has seen the construction of experimental instruments which in time may entirely replace the conventional transmission electron microscope.

5.2 BASIC PRINCIPLES

The principle of the SEM, used for examining a solid specimen in the emissive mode, is closely comparable to that of a closed-circuit TV system (Fig. 5.1). In the TV camera, light from the object forms an image on a special screen, and the signal from the screen depends on the image intensity at the point being scanned. The signal is used to modulate the brightness of a cathode-ray tube (CRT) display, and the original image is faithfully reproduced if (a) the camera and display rasters are geometrically similar and exactly in time, and (b) the time for signal collection and processing is short compared with the time for the scan move from one picture point to the next.

In the SEM the object itself is scanned with the electron beam and the electrons emitted from the surface are collected and amplified to form the video signal. The emission varies from point to point on the specimen surface and so an image is obtained. Many different specimen properties cause variations in electron emission and so, although information might be obtainable about all these properties, the images need interpreting with care. The resolving power of the instrument cannot be smaller than the diameter of the electron probe scanning across the specimen surface, and a small probe is obtained by the demagnification of the image of an electron source by means of electron lenses. The lenses are probe-forming rather than image-forming, and the magnification of the SEM image is determined by the ratio of the sizes of rasters scanned on the specimen surface and on the display screen.

For example, if the image on the CRT screen is 100 mm across, magnifications of $100 \times$ and $10\,000 \times$ are obtained by scanning areas on the specimen surface 1 mm and 10 μm across, respectively. One consequence is that high magnifications are easy to obtain with the SEM, while very low magnifications are difficult. This is because large-angle deflections are required which imply wide-bore scan coils and other problem parts, and it is more difficult to maintain scan linearity, spot focus and efficient electron collection at the extremes of the scan.

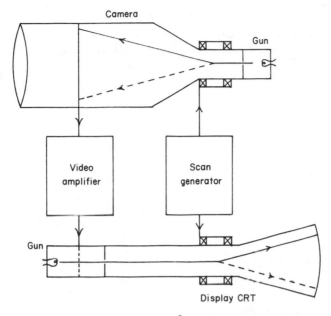

Fig. 5.1(a).   Closed-circuit TV.

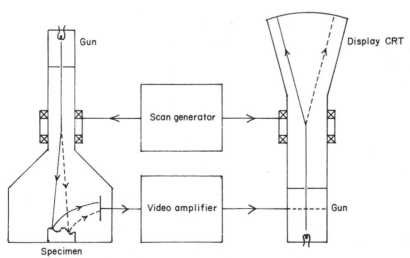

Fig. 5.1(b).   Scanning electron microscope.

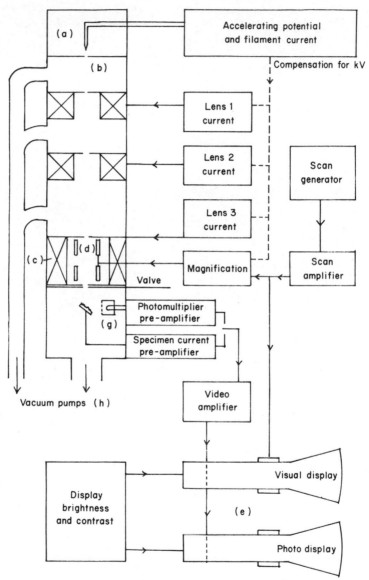

Fig. 5.2. Three-lens SEM (schematic).

5.3 INSTRUMENT DESIGN

Figure 5.2 shows the general layout of a typical three-lens SEM. Most of the features are common to all commercial SEMs.

(a) The electron source is a thermionic emission gun as used for the transmission electron microscope (TEM). The tungsten hairpin cathode gives a reasonable combination of brightness and lifetime when used in an easily maintained vacuum of 10^{-4} torr. Pointed filaments, lanthanum hexaboride (LaB_6) cathodes and cold-field emission tips are in use in some commercial instruments to give brighter emission. In general, the brighter the gun, the better the vacuum required for good lifetime and stability. Thus more complex pumping and isolation arrangements are needed for these brighter sources.

The accelerating potential of the gun can be varied typically in the range 1–50 kV (up to 1 MeV has been used experimentally).

In the absence of aberrations, brightness (units of $A\,m^{-2}\,sr^{-1}$) is a constant and is therefore the same at the specimen as it was at the gun. The tungsten thermionic emitter at 30 kV has a brightness (β) of about $4 \times 10^8 A\,m^{-2}\,sr^{-1}$ and so combinations of final probe diameter (d), current (i) and semi-angle of divergence (α) are possible such that

$$\beta = \frac{i}{d^2 . \pi\alpha^2} = 4 \times 10^8 A\,m^{-2}\,sr^{-1}$$

(b) The first image (cross-over) of the cathode formed by the thermionic gun is about 60 μm in diameter and this is demagnified by three magnetic lenses to a final diameter which may be smaller than 10 nm. Such a small probe may not be necessary for many applications, and the size may be altered by adjusting the first lens or lenses (numbering from the gun). The brighter types of gun give smaller first images and need less demagnification, so that two lenses or even one lens may be sufficient.

The focal lengths of the lenses depend on the energy of the electrons being focussed, and in order to keep the focal lengths constant as the accelerating potential is changed, it is usual to link the lens controllers to the EHT controller, so that lens current is reduced with accelerating potential.

(c) The final condenser lens (often and incorrectly called the objective) must have low aberrations if the smallest probe sizes are to be achieved. On the other hand, its magnetic field should not extend into the specimen chamber and it must operate at a focal length of 10 mm or more so that the specimen can be moved about and the emitted electrons collected; these features tend to go with increased aberrations. For the smallest probe sizes

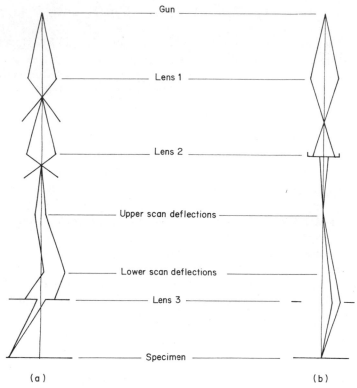

Fig. 5.3.   Schematic ray diagrams, (a) microscopy, (b) crystallographic contrast.

the lens aperture must be adjusted to the optimum size, typically around $100 \,\mu$m diameter. Since the probe is focussed from the aperture to the specimen surface, typically 15 mm away, the probe divergence is low (semi-angle $= 3 \cdot 3 \times 10^{-3}$ radian in this instance) so the depth of focus is large. This is one of the main reasons why the instrument is so useful: although most SEM micrographs are taken at magnifications within the reach of optical microscopes, there is far more information because more of the specimen is in focus.

For the highest-resolution STEMs (see Section 5.9), the lens is of the same type and quality as for the best TEMs. Similarly, the specimen is effectively inside the lens: size and movement are limited and the magnetic field near the specimen is relatively strong.

(d) The scan system must be compatible with the small final lens aperture, and there is not usually enough room to put deflection plates or coils between the lens and the specimen. A 'double-deflection' magnetic scan system is usually adopted, with the eight coils in the back bore of the final lens (Fig. 5.3(a)). The scan coils of the column are fed from the same scan generator, the 'magnification' control attenuating the current to the column while keeping the display on full width.

(e) For reasons to be explained, frame times of 10 or 100 s are often necessary to obtain good pictures. It is usual to adjust the microscope with a faster scan, then open the camera shutter and record a single slow frame.

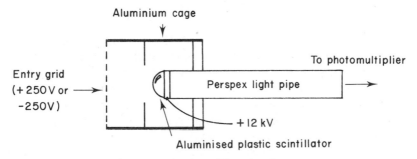

Fig. 5.4. Everhart–Thornley detector.

The visual examination can either be on the same short-persistence screen, if the visual scan is at normal TV speeds, or on a different long-persistence screen for frame times of the order of 1 s.

(f) It is normal for the specimen stage to permit the movement of the specimen in x, y and z, tilt and rotation while under examination, and special stages are available for heating, cooling, stressing and other experiments. In some SEMs the specimen is introduced to the chamber through an airlock, in others (e.g. the Stereoscan) the column is isolated from the chamber by a valve and the entire specimen chamber can then be let up to air. In a high-resolution STEM, the specimen stage resembles that of a TEM.

(g) The electron detector. For most purposes electrons emitted from the surface of a solid specimen are used to form the image. The electron detector has the task of detecting and amplifying this minute (10^{-12} A) electron current at high speed. The invention of a successful electron detector in 1960 by T. E. Everhart and R. F. M. Thornley contributed greatly to the success of the SEM concept. Figure 5.4 shows the construction. Electrons pass through a grid into a Faraday cage wherein lies a scintillator covered with a conducting layer and kept at a potential of

+ 12 kV. The flashes of light produced by the 12 keV (or more) electrons travel out of the vacuum along a light pipe and into a sensitive photomultiplier, which converts them back into an electrical signal. This signal represents a noise-free gain of 10^6 over the original electron flux, and the bandwidth is 10 MHz. The signal is now at a level to be amplified conventionally and modulate the display.

5.4   INTERACTIONS BETWEEN ELECTRON BEAM AND SPECIMEN

To understand the image formed in an SEM we consider the interactions between electron beam and the specimen surface. If the sample is insulating, charge might tend to build up and so the specimen is either coated with a thin conducting layer (e.g. evaporated gold) or the microscope is operated at a lower beam potential. For most insulating materials, the secondary emission increases as the incident electron energy is decreased down to 1 keV or so, and thus we can decrease the potential until no charging occurs. The disadvantage of working at low potentials is that gun brightness is reduced, giving a general loss of SEM performance. Therefore coating of insulating specimens is common practice. The gold (or carbon for X-ray analysis) can be evaporated or sputtered, the former needing rotary jigs for rough specimens.

The incident electrons can suffer scattering tending to reverse their original direction and give a flux of 'reflected' electrons; cause the emission of low-energy secondary electrons; produce photons of various wavelengths; contribute to the current flowing between specimen and earth; produce radiation damage; and heat the specimen. The last two effects may mean that the specimen is damaged by the examination, and that the micrograph must be taken quickly. The other effects are of interest for image formation; we can use a suitable detector and obtain a video signal from any of the electron or photon emissions, each giving us different information about the specimen.

Figure 5.5 shows a typical distribution of the energy of electrons emerging from the surface of a flat metal sample under a normally incident beam of energy several keV. Two distinct groups occur, separated into emitted and reflected electrons by an arbitrary dividing line at 50 eV. The total electron flux emerging from the specimen is usually between 0·1 and 0·9 of the incident beam current, i.e. it is of the order of 10^{-12} A under high-resolution conditions. If the grid on the Everhart–Thornley electron

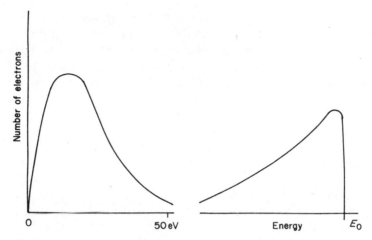

Fig. 5.5. Electron energy distribution emerging from flat specimen.

detector is biased to $+250$ V, secondary electrons are collected whatever their original trajectory from the specimen. If a bias of -250 V is used, the secondary electrons are repelled and the reflected electrons form the signal. In this case only those originally travelling towards the cage will be collected, so even though there are often more reflected than emitted electrons, the reflective signal from such a detector is usually very much weaker than the emissive. The scintillator and light pipe (without the cage) are used below the specimen for the transmission mode.

5.5 MODES OF OPERATION

The different ways of obtaining the video signal are the different modes of operation of the SEM. These are:

(a) Emissive: secondary electrons are detected. The Everhart–Thornley detector is used with positive bias.

(b) Reflective: back scattered electrons are detected. The Everhart–Thornley detector with negative bias is normally used, but a greater solid angle of collection can be obtained from annular detectors such as the semiconductor or channel-plate.

(c) Transmission: the electrons passing through a thin specimen are detected (more of this in Section 5.9).

(d) Absorptive: the current flowing between sample and earth is detected.

A special FET-input amplifier is normally used. In general the absorptive signal is complementary to emissive + reflective. It gives no new information and one loses the benefit of the sensitive detector. For some applications there is no other way of obtaining the signal.

(e) Conductive: an external e.m.f. is applied across part of the specimen and the current flowing in this external circuit is used. This can give useful information about semi-conductor materials.

(f) Luminescent: visible, infra-red or ultra-violet radiation is detected. In order to eliminate the other signals, it has become usual to collect the light with an ellipsoidal mirror, the specimen at one focus and a light-pipe at the other. At the far end of the light-pipe, the light can be detected with a photomultiplier, with or without spectral analysis.

(g) X-ray: the X-radiation is detected. By using a dispersive detector, elemental microanalysis can be performed.

(h) Auger: the Auger electrons are a special group of secondary electrons with characteristic energies in the range 0–2 keV. By energy spectrometry of these electrons, elemental analysis can be performed. However, the relatively low energy implies a low escape depth, so that only the first few nanometres of the surface are analysed. Unless the chamber vacuum is around 10^{-9} torr, unwanted surface layers are at least of this thickness. Although Auger potentially offers a means of analysing the light elements (difficult with X-rays), it is a specialised technique in itself and not at present a practical mode of SEM operation.

(i) Energy-loss spectroscopy: a corresponding group of backscattered electrons exhibit small energy losses from the original beam energy. Spectroscopy of these can also be used for surface analysis. The technique is in its infancy and is mentioned here for the sake of completeness.

5.6   CONTRAST MATRIX

Different features of the specimen give rise to intensity variations in the signals, thus forming the image. Different contrast mechanisms affect the signals in the different modes of operation, and Table 5.1 is an attempt at a complete summary. To understand the table, refer first to the emissive mode and read the following description of the contrast mechanisms; this should clarify the entries for other modes.

(a) Topographic. Figure 5.6 shows how the emissive signal rises steeply with the angle between the surface normal and the incident beam direction, for angles towards the collector. The signal also increases for tilts in other

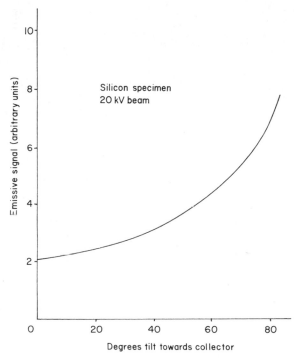

Fig. 5.6.   Effect of surface tilt on emissive signal.

directions, but less steeply, due to losses of electrons to the pole-piece and parts of the stage. A rough specimen thus appears as if viewed from the direction of the incident beam, while being illuminated predominantly from the electron collector. This means that rough specimens have a realistic 'shadowed' image which is easy to interpret, and is another of the main reasons why the SEM has proved valuable.

(b) Atomic Number. A specimen which is smooth but has areas of different materials exhibits contrast in the emissive mode because the signal rises with Z up to about $Z = 40$.

(c) Surface Potential. A smooth specimen of homogeneous material shows contrast if there are voltage differences (e.g. across a p–n junction). This is due to a variation in emissive signal of about 5% per volt.

(d) Magnetic Fields. The presence of surface magnetic fields alters the trajectory of emitted electrons. This is known as Type I magnetic contrast. Domains in cobalt can be revealed with a contrast of about 30%.

TABLE 5.1. *Contrast Matrix*

(The effect of various specimen features on the signal in various SEM modes of operation. The maximum value of $C = \delta S/S$ is shown where known. √ indicates an unquantified effect. x indicates no effect detected)

Mode \ Feature	Emissive	Reflective	Transmission	Absorptive	Conductive	Luminescent	X-ray
Topography	Dependent on angle C up to 2	Dependent on angle C up to 2	Affected by thickness	Dependent on angle C up to 2	x	√	√
Atomic Number	C approximately proportional to Z	C approximately proportional to Z	√	C approximately proportional to Z	affected by defects	molecular effect on λ	λ depends on Z
Potential	C approximately 5%/volt	√	—	√	affected by internal fields	—	—
Magnetic Field	Type I: C up to 0·3 Type II: C up to 0·001	Type II only C up to 0·001	Lorentz contrast	Type II only C up to 0·001	—	—	—
Crystallography	C up to 0·1.	C up to 0·1	Diffraction and Kikuchi contrast	C up to 0·1	—	—	Kossel lines

Fig. 5.7. Absorptive mode, 30 kV. Annealed Cu specimen, chemically polished.
Crystallographic contrast (grains).

(e) Crystallographic. If the specimen is flat, homogeneous, free of surface potential differences and non-magnetic, we can still see contrast because of the different crystal orientations of individual grains (see Fig. 5.7). The maximum variation to be expected is about 5%. If the specimen is a large single crystal, and is examined at low magnification, we obtain the electron channelling pattern (ECP) (Fig. 5.8). This arises because of the variations in angle between the incident beam direction and the crystal lattice planes in the course of a frame. Since the signal depends only on the direction of incidence, the ECP remains stationary when the crystal is moved about (without tilting) under the beam. Equally, the sharpness of the pattern does not depend on the spot size but on probe divergence. The ECP, which resembles a Kikuchi pattern, can be used to orient, and sometimes to identify, the crystal lattice being examined (see Section 5.8).

When Table 5.1, the contrast matrix, was first drawn up, it was mostly blank and even now there are many gaps and uncertainties. These indicate fields for valuable research, and in the future we can look forward to filling these spaces with firm answers as a result of such research.

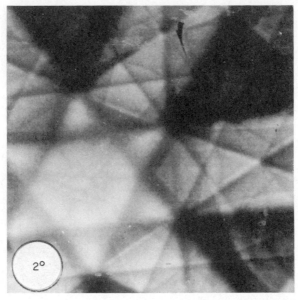

Fig. 5.8.   Reflective mode, 20 kV. Single-crystal Si specimen. Crystallographic
contrast (electron channelling pattern—ECP).

5.7   VISIBILITY OF FEATURES

5.7.1 Signal-to-noise Ratio

It is important to note the signal variation to be expected from each contrast
mechanism, because this determines whether the feature concerned will be
visible on the screen. Even if no noise is introduced by the amplifiers, the
video signal will inevitably have a basic noise level. This is because the
emission of electrons from a hot filament (or any other gun) is a random
process and subject to statistical variations. This basic noise level cannot be
reduced, and may be increased by the processes of electron emission from
the specimen, electron detection and amplification. The result is that a
variation δS in a signal of general level S may be partially obscured by a
noise level δN (Fig. 5.9). The criterion for visibility often adopted is $\delta S >
5\delta N$. Let the number of electrons striking a given point on the specimen
surface in the course of a frame of t seconds be n. The statistical variation in
n is $n^{1/2}$, so that the basic signal-to-noise ratio cannot be better than
$n/n^{1/2} = n^{1/2}$.

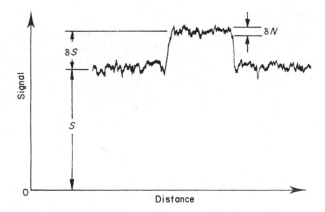

Fig. 5.9.   Variation δS in signal of level S with noise δN.

Thus

$$\frac{\delta S}{S} > 5\frac{\delta N}{S} \quad \text{and} \quad \frac{S}{\delta N} = n^{1/2}$$

The signal variation $\delta S/S$ we call the contrast C. If we are trying to see this on the screen then the number of electrons striking each point on the specimen surface in the course of a frame must be greater than $25C^{-2}$, neglecting any additional noise from specimen, detector, amplifier, etc.

Now $n = it/L^2 e$, where t = frame time, i = probe current, L^2 = number of picture points per frame, and e = electronic charge; and in turn the probe current $i \sim 2\cdot 5\,\beta d^2\,\alpha^2$, where β = gun brightness, d = gaussian spot diameter, and α = semi-angle of divergence; and we have said that to make a contrast level C visible $n > 25C^{-2}$.

Some consequences of this are:

(a)   If the contrast is low, then a high probe current, or long frame time, will be necessary for a given visibility above the noise.

(b)   A high probe current implies a large spot or high beam divergence (i.e. large final aperture, if the beam is focussed).

(c)   Attainable resolution depends on contrast. Resolution test specimens are therefore usually chosen to have a strong topographic contrast ($C = 2$) on the required scale.

(d)   A high-brightness gun can improve the resolution (provided we are not aberration limited) or decrease the frame time for a given resolution.

5.7.2 Highest Resolution

A specimen may have a signal variation of 2:1 due to topographic contrast, and for $C = 1$, $n > 25$ and for $t = 100$ s and $L = 625$, $i > 1.6 \times 10^{-14}$ Å.

Thus with a tungsten gun with $\beta = 4 \times 10^8$, and a divergence $\alpha = 3 \times 10^{-3}$ radians, $d > 1.3$ nm.

At this level the lens aberrations and stability of the specimen stage will be extremely important. The design of normal SEM lenses means that aberrations are in fact limiting and we can obtain a minimum usable spot size of around 3 nm.

The minimum usable spot size may be equated to the best resolution obtainable in the transmission mode. This would also apply to solid specimens examined in the emissive mode if the signal were obtained solely from those secondary electrons generated by the direct action of the primary beam. In fact, secondary electrons are produced also by emerging backscattered electrons and these may have originated $0.5 \, \mu m$ or more away from the point of incidence of the primary beam. These 'indirect' secondaries may be regarded as contributing to the noise, rather than to the signal, and so the resolution may be degraded. This is known as the 'spreading effect'. In the case of the reflective and X-ray modes, the effect is more serious and may be the most important factor limiting the resolution. For solid specimens examined in the emissive mode, 5 nm is about the best resolution expected. By using a special lens and LaB_6 gun, A. N. Broers has demonstrated 2.5 nm in the emissive mode.

In the transmission mode, the final lens resembles that of a TEM so that aberrations can be reduced. It is then worth using a higher-brightness gun so that the lens aberrations are still limiting and the ultimate performance can be achieved. The resolution of a field emission STEM can be as good as, or better than, that of any TEM.

5.8   SELECTED AREA DIFFRACTION

5.8.1 Solid Specimens

Since the electrons emerging from the surface of a solid specimen have all suffered inelastic scattering, no diffraction spots will be seen. The reflection Kikuchi pattern can be seen by having a stationary probe and putting a fluorescent screen near to the specimen. Alternatively, if the beam scans through appreciable angles in the course of a frame, we can use the normal electron detector and obtain the ECP, the reciprocal analogue of the reflection Kikuchi pattern. The ECP can be obtained from a small selected area if the scan is arranged to change the direction of incidence of the probe while it always strikes the same small area on the specimen surface. Each point on the display screen then corresponds to a particular direction of incidence rather than to a particular point on the specimen. This is normally achieved by a single scan deflection in conjunction with the final lens, the lens being used not only to focus the probe but to bring the scan back to the same point on the specimen in the required manner (Fig. 5.3(b)). The second lens must form its image in the plane of the scan deflection, and this may involve a much lower lens current than usual. To keep the probe divergence low (required for a sharp ECP) a small aperture is introduced into the column in the region of lens 2. Finally, the lens 3 aperture must be enlarged to include the angular scan. A total scan angle of more than 8° is usually necessary if the ECP is to be identified. A lens 2 aperture reduced from 1 mm to 100 μm, and a final lens aperture enlarged from 100 μm to 3 mm, are typical. In order to display the ECP, which is of low (up to 10%) contrast, a high beam current and hence a large spot size is necessary. Lens 1 is adjusted for the required conditions, and a 2 μm probe is usually satisfactory. The specimen is examined with this probe at a magnification of 200 × or 500 × (double-deflection scan), and the grain of interest brought

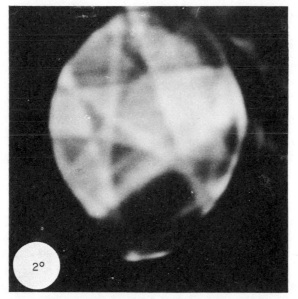

Fig. 5.10.   Absorptive mode, 30 kV. Grain A of Fig. 5.7. Crystallographic contrast (selected area channelling pattern—SACP).

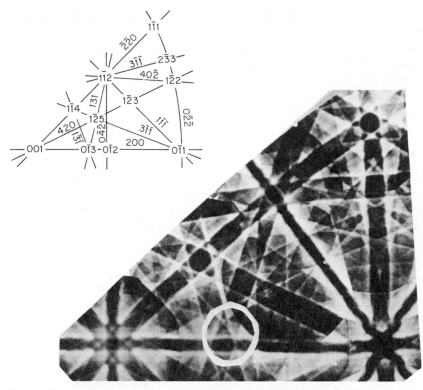

Fig. 5.11. Absorptive mode, 30 kV. Single-crystal Cu specimen. Montage of 9 ECPs to form a complete channelling map. Direction of Fig. 5.9 identified.

to the centre of the screen. On switching to single scan and $20 \times$ 'magnification' the ECP appears and may be photographed (Fig. 5.10). Since a very wide aperture is used in the final lens, the minimum diameter of selected area from which an ECP can be obtained is limited by spherical aberration. The minimum diameter is $d_{min} = \frac{1}{2}C_s\theta^3$ where θ is the semi-angle of scan. C_s is markedly reduced if the lens is operated at short focal length, and for $C_s = 30$ mm, $\theta = 4°$ we obtain $d_{min} = 5\,\mu$m. The ECP can sometimes be identified analytically by measuring the angular widths of the bands (equal to twice the Bragg angle for the planes concerned) but indexing is most easy by comparison with a standard map of all possible orientations, prepared for the material and accelerating potential in use (Fig. 5.11). The ECP of Fig. 5.10, which was obtained from grain A of Fig. 5.7, is identified on the map.

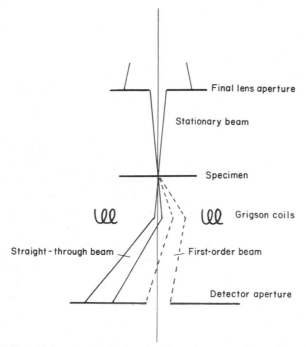

Fig. 5.12. Grigson method of selected area diffraction for transmission specimens. The first-order diffracted beam is being scanned over the detector.

5.8.2 Thin Specimens

In the transmission mode, diffraction pattern or transmission Kikuchi pattern can be seen depending on specimen thickness. Again, the beam can be stationary or rocking. If the beam is stationary, then the pattern can appear on the normal display screen if there are scan deflection coils (Grigson coils) between specimen and detector (Fig. 5.12). If the beam is rocking, then the normal detector can be used without further scan coils. In either case, the sharpness of the pattern depends on the probe divergence and on the angle subtended by the detector aperture at the specimen.

The size of the small selected area is normally fixed, as for solid specimens, by the area illuminated by the probe. However, if there are projector lenses between specimen and detector then a smaller area may be selected by imaging the specimen on to the detector final aperture.

The techniques are still not fully developed, a recent suggestion being a 'double rocking' technique with synchronised double deflection on both sides of the specimen. The beam is incident on one small area of the specimen at all times. Together with variations of the beam divergence, collected angle and selected area, such scanning methods are the subject of important research.

5.9   SCANNING TRANSMISSION ELECTRON MICROSCOPY (STEM)

5.9.1 Using a Normal SEM

Here we are using the transmission mode of operation of a normal SEM. An electron detector (normally the scintillator and light pipe, without the cage) is placed on the far side of the specimen from the final lens. The aperture of the detector (subtending a semi-angle β_s at the specimen) has an important influence on the information collected. Electrons which have suffered little scattering (elastic and low-loss electrons) will be within a cone of angle approximately equal to the incident beam divergence. Outside this will be the inelastically scattered electrons. Different features of the specimen affect the relative intensities of the elastic and inelastic signals. Thus, we would like:

(a) to be able to separate the elastic and inelastic signals

(b) to vary β_s depending on the incident beam angle and the contrast required.

In a SEM conversion, this can be achieved by having a range of apertures available to cover the detector disc, together with a stop for the elastic signal. Unfortunately the apertures also act as field stops, so that the field of view on the specimen is severely limited. The usable area is less than that of the aperture because the beam must not approach the shadow edge—a partial loss of the elastic signal would occur.

Thus serious quantitative STEM work is difficult in a converted SEM. However, within the limitations one can perform useful microscopy. The greatest advantages are:

(a) High resolution. Since the specimen need no longer be seen by the emissive detector, the final lens can be operated at the highest possible excitation. This dramatically reduces C_s, enabling the best possible resolution to be achieved—an instrument capable of 6 nm in the emissive mode will probably resolve 3 nm in transmission.

(b) Diffraction. This has already been mentioned in the last section (5.8.2). Extra coils are needed for the Grigson technique, but this is then somewhat easier to use than the rocking-beam method.

(c) X-ray analysis. Since the specimen is still relatively exposed, the use of X-ray detectors is far easier than in a TEM or purpose-built STEM.

(d) Thick specimens. In general, specimens prepared as for normal TEM will give good results at the lower SEM voltages, and some results can be obtained for thicknesses of 1 μm or more.

5.9.2 Using a Normal TEM

The presence of imaging lenses between specimen and detector means that β_s can be varied as desired and need not be field-limiting (the aperture can be arranged to lie in the back focal plane of a post-specimen lens). Furthermore, the low-aberration objective lens means that a better resolution can be obtained than in the SEM. However, the resolution is not as good as in a normal TEM and this is because the smallest probe sizes are not usable: there is not enough beam current. Referring back to Section 5.7.2, we cannot expect better than 1·3 nm with a 30 kV tungsten gun, and this figure assumes a high contrast of $C = 2$ and no limits from aberrations. At 80 to 100 kV the likely lower limit is 1·0 nm.

The advantages of this mode of operation are:

(a) Observation of thicker specimens at a given accelerating voltage. For amorphous specimens at least, the loss in resolution with thickness is not as serious in STEM as in TEM. The limitation with STEM is caused by the beam broadening, and in TEM by plural scattering in the specimen and also by chromatic aberration: electrons which have lost significant energy are not imaged. If we could get sufficient current into the spot then the STEM could give approximately equivalent resolution to the STEM for specimens about five times thicker. In practice, when using a converted tungsten-filament TEM, the resolution in the STEM mode cannot be better than about 1·0 nm from signal-to-noise. Thus there can be a significant advantage for the tungsten-filament STEM when viewing specimens around 1 micron thick (operating at 100 kV).

(b) Microdiffraction. By using a Grigson or rocking-beam method, diffraction patterns of some sort can be obtained from regions 10 nm across. However, the resolution and quality of these patterns are usually greatly inferior to normal SAD patterns obtained from micron-sized regions in the normal TEM.

5.9.3 Using a Special-purpose Instrument

Based on the discussion above, we can see that a purpose-built STEM would have:

(a) high-brightness gun and
(b) lenses between specimen and detector.

A final desirable refinement is to add an energy-selecting detector so that energy-loss spectrometry can be performed, or energy-selected image displayed.

At present, experimental microscopes have been made with all three features, but current commercial instruments lack the imaging lenses.

The most striking STEM results have undoubtedly been those of A. V. Crewe and his colleagues, using instruments with field-emission guns and electron spectrometers. The accelerating voltages have been in the range up to 100 kV. With these instruments the imaging of single heavy-metal atoms on thin substrates has been demonstrated. The lightest element so far imaged and identified is Ag. Vacuum Generators Ltd use a similar arrangement in their HB5 STEM instrument, examples of which are being evaluated for both materials and biological studies.

The resolution of such a field-emission STEM is limited by the same factors as the normal TEM. Advantages of the method include:

(a) Imaging of thick specimens. Experimental instruments operating up to 1 MeV are nearing completion: these should enable the thickest specimens yet to be viewed.

(b) Contrast can be improved for certain specimens. This is a consequence of being able to separate the elastic and inelastic electrons and derive signals based on the ratio of the two intensities.

(c) The third and most exciting advantage, not yet proved experimentally, is the possibility of correcting all the third-order aberrations simultaneously. In a TEM this is theoretically impossible, but the limitation is removed in the STEM. We look forward therefore to STEM instruments with a substantially better performance than any TEM: such instruments may indeed replace the TEM entirely.

BIBLIOGRAPHY

Rather than cite individual papers, I refer the reader to two sources containing the latest research in the subject:

Annual SEM Symposia Proceedings (11 vols., 1968 to 1978) obtainable from:

SEM Symposia Office, Metals Research Division, IIT Research Institute, Chicago, Illinois 60616, USA

Electron Microscopy and Analysis Group (EMAG) Proceedings (4 vols., 1971, 1973, 1975 and 1977) obtainable from:

The Institute of Physics, Techno House, Redcliffe Way, Bristol BS1 6NX, UK

6

Instrumentation in Electron Probe Microanalysis

W. J. M. SALTER

Iron and Steel Industry Training Board, Sheffield, UK

6.1 INTRODUCTION

In discussing instrumentation for electron probe microanalysis (EPMA) it is not proposed that the electronics will be considered in any great detail, but only in so far as they are the means to control the system and as they effect repeatability, reproducibility and stability.

The broad requirements are shown in the flow diagram in Fig. 6.1. The EHT supply to the gun generates electrons from the cathode at potentials which can be varied normally between 0 and 60 kV—typically 25 kV. These electrons enter the probe-forming electron-optical system, consisting basically of two electrostatic or electromagnetic lenses, which produces a finely focussed beam of electrons on the sample surface. Within the electron-optical system are situated two pairs of scanning coils, which allow the beam to be systematically scanned over a region of interest on the specimen surface. Chapter 1 contains a description of electron-optical systems, and hence this will not be considered further.

The electron probes on early instruments were typically 1 μm resolution (10 000 Å), but these are now generally between 500 and 1000 Å and in the case of scanning electron microscopy (SEM), which is frequently combined with microanalysis, the electron resolution is typically 100 Å.

The specimen chamber is a very congested area of the instrument because it is necessary to move the sample mechanically in the X, Y and Z directions, to accommodate air locks for specimen change, reference standards, optical microscope path, reflected and absorbed electron detection, usually two X-ray paths to the spectrometers and the facility to carry out energy dispersive analysis and secondary electron detection for SEM. Hence the design of the specimen chamber, and particularly of the objective lens is very important, in order to accommodate these required facilities.

The EPMA operator requires information from the specimens being

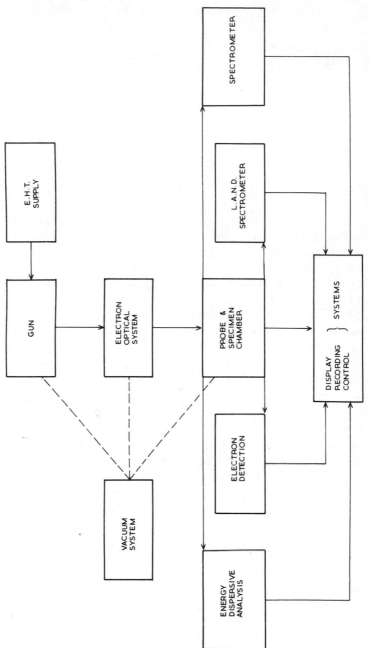

Fig. 6.1.   Schematic requirements for electron probe microanalysis. L.A.N.D. = low atomic number device.

analysed, which is supplied both by electrons and X-rays. In the case of electrons, information is obtained in three main ways: (a) high-energy electrons backscattered from the specimen surfaces—these being affected both by topography and variations in atomic number; (b) absorbed incident electrons, giving rise to varying specimen currents, which are also modified by topography and atomic number; and (c) secondary, emitted electrons, which are the principal means of performing SEM. All these modes give valuable information about the specimen and additionally give the operator the means to locate specific features.

Microanalysis relies on the emission of X-rays, details of which are discussed in Chapter 7. Hence, the reliable detection and recording of the X-rays generated are of great importance. There are two options open.

(a) Wave-dispersive analysis (WDX) is the traditional EPMA technique, requiring a crystal spectrometer system in order to diffract and separate the X-rays of various wavelengths, which are subsequently detected by a proportional counter and fed into the recording and display systems.

(b) More recently, energy-dispersive analysis (EDX) has gained momentum and the spectrum is based on the varying energies of the X-rays rather than their wavelengths. This technique has particular advantages in SEM and for quick elemental identification in EPMA. However, the ability to use energy-dispersive analysis quantitatively is being developed with some success.

The recording, display and control systems are a combination of electronic and mechanical configurations. It is important that the instrument should be easy to operate and be as 'foolproof' as possible, commensurate with the complexity of the technique. The operator will therefore require a 'comfortable' layout of the controls to avoid excessive fatigue, and to be able readily to gather all the information available to complete an examination successfully.

The final aspect which must be considered is the vacuum system. Typically the complete probe-forming system, specimen chamber and spectrometers need a vacuum of the order to 10^{-4}–10^{-5} torr. Into this vacuum many controls have to pass, thus necessitating careful design of the system in order to ensure vacuum-tight seals, efficient pumping lines and safety systems to avoid accidental air admittance.

6.2   ELECTRON SOURCES

Whilst the electron-optical systems have been examined in some detail, it is important to appreciate that there are more electron sources available now

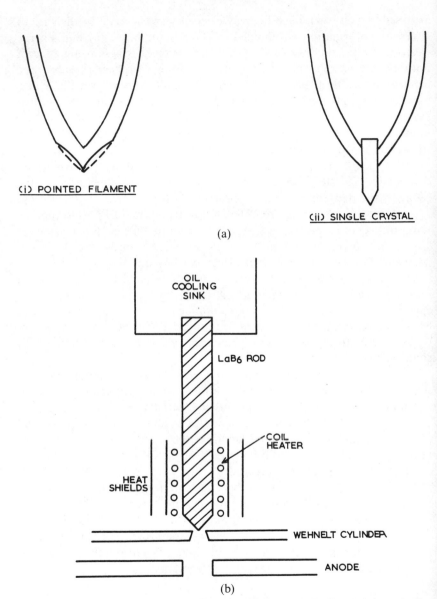

Fig. 6.2. (a) Schematic representation of (i) pointed W-filament by grinding (ii) single-crystal W-wire etched to a point and welded to the hairpin. (b) Schematic diagram of LaB$_6$ source.

than just the directly heated W-wire hairpin filament. Table 6.1[1,2] summarises the sources either available or in development. They fall into three categories: (a) directly heated thermionic emitters; (b) indirectly heated thermionic emitters; (c) field emitters.

The conventional W-filament source has the advantages of requiring only a moderately good vacuum (better than 10^{-4} torr), and of being cheap and well understood. However, in order to have improved resolution in microanalysis, consistent with good X-ray intensity generated from the specimen, electron sources much brighter than the conventional ones are necessary. Probably the indirectly heated sources offer the best option, because little instrument modification is required and a vacuum improvement to 10^{-6}–10^{-7} torr only is necessary in the gun region.

Figure 6.2 shows the schematic configuration of three electron sources and Fig. 6.3 compares probe current versus probe radius for new and conventional electron sources.[3]

Clearly, for those instruments requiring high-resolution electron beams,

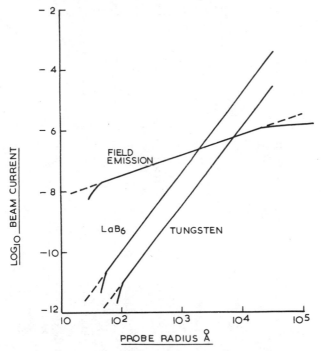

Fig. 6.3.   Variation in probe current as a function of probe radius for conventional W-filaments, indirectly heated sources (LaB$_6$) and cold-field emission.

TABLE 6.1
Typical Data Associated with Electron Guns

Source	Relative brightness[a]	Source size	Stability	Life	Vacuum required	Instrument suitability
Directly heated						
Tungsten hairpin	1	$50\ \mu m$	$<1\%$	50h	$<10^{-3}$ torr	TEM EMMA EPMA SEM STEM
Pointed tungsten filament	2–10	$10\ \mu m$	$\sim 3\%$ (mainly drift)	20h	$<10^{-4}$ torr	As above
Modified Wehnelt	5–10	Probably better than $10\ \mu m$	$<1\%$	100h	$<10^{-3}$ torr	As above
Indirectly heated						
LaB_6	30	$1\ \mu m$	$<1\%$	300h	10^{-6} torr	EPMA SEM STEM
Field Emitters						
Cold-field emission	500	$50\ \text{Å}$	$3–5\%$	Varies with vacuum	10^{-10} torr	SEM STEM
Hot-field emission	500	$50\ \text{Å}$	$<5\%$	100h	10^{-7} torr	High resolution SEM STEM

[a] Related to conventional tungsten hairpin filament.

field emission potentially gives a much brighter source, but a vacuum approximately 10^{-10} torr is usually necessary and X-ray emission is low. For most electron probe microanalyser systems, with a probe radius approximately 10^3 Å, the practical brightness improvement of LaB_6 for relatively little instrument modification seems the most promising, and is in fact now becoming commercially available.

From time to time the gun filament or electron source will require changing. With the conventional hairpin W-filament this is fairly frequent and hence a gun vacuum isolation valve is required in order to confine the necessary vacuum loss solely to the electron gun.

The problems of quantitative analysis are examined in detail in the next chapter, but it will be appreciated that a sufficient number of X-rays must be recorded in a suitably convenient period of time so that statistically meaningful analyses can be obtained. The brightness and diameter of the electron probe are two limiting factors, particularly as there is an increasing requirement to combine EPMA and SEM; thus, higher brightness sources are essential to allow for the need to decrease probe diameters.

6.3   SPECIMEN CHAMBER

Having obtained an electron probe of suitable resolution and brightness, it is in the specimen chamber that the probe reacts with the sample generating a range of X-rays and electrons of various energies. The following sections will consider in more detail what needs to be fed into the specimen chamber and what data need to be extracted from it.

6.3.1   Optical Viewing of the Specimen

In EPMA some form of optical viewing system is necessary in order to locate the feature of interest on the specimen. Various commercial instruments have different designs, but essentially, in the congested region of the specimen chamber, an optical path needs to be found. In the SEM mode this need does not exist generally because a different type of examination is carried out and this, coupled with the low magnification made available by the electron detection system, largely removes the need. It is not difficult to construct a system to view the specimen either optically or by the electron beam. Much greater complication occurs when simultaneous optical and electron beam viewing is required—this usually necessitates oblique optical viewing. The optical microscope needs to have good resolution, probably offering two or three magnifications in the range

$100 \times$ to $500 \times$, and to be so aligned that the feature of interest is viewed either simultaneously or, alternatively, with the electron probe, without the need to move the specimen, i.e. the fields of view must be concentric.

A number of instruments have both alternative and simultaneous optical viewing. The advantage of the latter is that it allows cathode luminescence to be observed, i.e. that feature which in certain cases gives rise to optical coloration when the electron beam reacts with the specimen. This effect is reported by many workers, and particularly in the oxide systems the visual coloration gives an indication as to the general composition, e.g.

Corundum—Al_2O_3	Red
Crystabalite—SiO_2	Light blue
Mullite—$3Al_2O_3 \cdot 2SiO_2$	Blue–white
Spinel—$MgO \cdot Al_2O_3$	Green

For the microanalyst using thin biological or ceramic/refractory specimens then the facility to view optically both in reflection and in transmission is a distinct advantage.

6.3.2 Specimen Holder and Movement

In most conventional microanalysers the incident electron beam is normal to the surface of the specimen, and the specimen movement is confined to the X and Y directions for feature location, with a more restricted Z movement for optical and electron beam focussing. Certain instruments also have the facility to rotate the specimen about the Z direction. In SEM examinations the ability to tilt becomes necessary. It will generally be found that with an SEM the optimum position for carrying out scanning electron microscopy is not the same as for analysis, particularly WDX, where the Rowland geometry of the spectrometer must be satisfied (see Section 6.5). Thus, in these circumstances the specimen must be brought to a standard position to ensure that the angles in relation to the incident probe and the crystal spectrometer are correct.

In the conventional microanalyser, focussing the specimen optically using the Z movement with fine adjustment to the electron probe satisfies the analytical geometry.

The ability to rotate the specimen about the Z axis, together with complete X and Y direction coverage offers particular advantages when carrying out segregation scans across features. In these examinations the normal procedure is to drive the specimen under a static probe, using accurately variable servo-motors, in either the X or the Y direction. Clearly, if the required direction is not parallel to either X or Y then rotation would seem necessary. However, a rotational facility frequently limits the specimen size to a maximum of about 40 mm. Quite often the microanalyst wishes to examine larger specimens. In those cases where a larger specimen is used, e.g. 90 × 50 mm, then designers have achieved a range of oblique scanning directions by using simultaneously the X and Y servo-motor drives, balanced to give angles of inclination to, say, the Y direction at 15° intervals. In deciding what is required, therefore, it is necessary to determine whether it is more important to be able either to rotate specimens or to examine larger specimens. The need both to rotate and tilt in SEM, whether in conjunction with analytical facilities or not, will limit the specimen size.

For EPMA, as opposed to SEM, it is necessary to have standards of various elements or alloys present in the chamber, together with the specimen. The specimen holder therefore must be designed to accommodate this requirement. The X and Y drives to the specimen holder, whether mechanical or manual, need to indicate the co-ordinates of any particular region of the sample, or the position of the standards, for future reference. Most microanalysers have available 'push-button' systems for location of standards plus several variable dials, which allow the operator to pre-set positions on the specimen and return to these automatically, utilizing the push-button system.

There are cases where segregation on a macro scale (e.g. several millimetres or centimetres) requires determination. Often in these instances, a continuous segregation profile is not necessary and indeed would be time-wasting. A useful addition is to have a step scanning system which, if the instrument utilizes some form of automatic record and print-out system, will move the specimen along the required line of scan carrying out analyses at pre-set intervals, which can be variable according to the requirements of the investigation. These intervals may be variable between, for example, steps of 100 μm to several millimetres.

6.3.3 Specimens and Standards
Microanalysis is carried out on a variety of specimens, including metallic, ceramic, organic and inorganic materials. The condition of the specimens and standards can markedly affect the stability and sensitivity of the instrument. The incident electron beam reacts with the specimen, such that part of it is backscattered and part absorbed, depending generally upon the mean atomic number and topography of the region under the probe (see Section 6.4). The instrumental stability can be severely affected if that part of the electron beam that is absorbed cannot be conducted away. Indeed the

current density within the specimen close to the incident probe can be very high. Thus the feature being analysed must be capable of conducting the electrons to avoid 'charging up' and hence special preparation techniques are needed for non-metallic materials. The microanalyst will generally require subsidiary equipment which enables him to vacuum-deposit a conductive coating on to the surface of the specimen and standards if necessary. There are several commercial coating units of this type, generally used for obtaining carbon replicas for conventional electron microscopy. The method of evaporation depends upon the coating material required, but generally, except for carbon, will utilize a resistance-heated boat; alternatively an electron bombardment technique may be used. Typical coating materials are carbon, aluminium, copper and gold. Beryllium is sometimes used, particularly when it is necessary to detect very soft X-rays, but this element needs some care because of its potential health hazards.

As the operator will not wish to recoat his standards with every new specimen, because this involves complete repreparation, some system of accurately monitoring the coating to give consistent thickness every time is necessary. Two systems are commonly used:

(a) The specimen(s) in the evacuated chamber are placed a predetermined distance from the evaporation source, and a known weight of the coating material is totally evaporated.

(b) The specimen(s) are similarly located a predetermined distance from the evaporation source, but a double quartz oscillator monitor is used to register the film deposition. In this case the active quartz crystal in the vacuum chamber changes its oscillation frequency as the deposition progresses, in relation to the reference quartz crystal. This difference is displayed on a meter and when a predetermined frequency change has occurred the coating thickness required has been achieved.

As with the gun filament, but in this case several times each day, the specimen and standards will be changed. Thus, it is again necessary to incorporate a specimen chamber vacuum isolation valve, so that the minimum vacuum is lost during a specimen change.

Whilst vacuum systems appropriate to these types of instruments are discussed in Section 6.9, several phenomena can occur in the specimen chamber, at least partially attributable to the vacuum condition, which can cause erroneous analyses to occur. The most common of these is attributed to the dissociation of the hydrocarbon vapours at the specimen surface under the action of the electron beam, giving rise to the carbon contamination marks so familiar to microanalysts. Generally, the poorer the vacuum, the more rapidly this contamination will occur, resulting in a decrease in the measured X-ray intensity, particularly important with the lighter elements and their soft X-rays. This effect will be examined further when discussing instrument accuracy. These hydrocarbons can come from several sources, but the most common are from oils and greases used in the vacuum system and from the specimen/standard itself.

6.4 ELECTRON DETECTION

Having described the requirements of the specimen chamber and what needs to be fed into it, it is now equally important to extract information in order to carry out the investigation and required analyses. The use of the electrons backscattered, absorbed or generated as secondary electrons by the specimen plays an important part in microanalysis, allowing the operator to view a sample that can give an appearance similar to that seen optically. Additionally, by using the various generated sources, electronically modifying and combining them in various ways by selective detection and apportionment, a variety of effects can be achieved. The scanning electron microscope of course makes greatest use of these facilities and—as the user will discover—because of the compatibility of SEM and EPMA, several manufacturers offer a combined instrument with considerable success, although perhaps some of the ease of analysis is lost in comparison with the modern microanalyser.

6.4.1 Backscattered Electron Detection

Two basic systems are available to detect the high-energy backscattered electrons. The second of these will be found in modern instruments and is a more reliable system:

(a) The use of a phosphor, which scintillated when bombarded by electrons and was sited on the end of a light guide, was normal in early microanalysers. The light guide fed a photomultiplier tube, from which the signals were detected by a scintillation counter. Amplifiers and appropriate electronics fed the variable impulse to a display screen which scanned in synchronization with the electron probe—Fig. 6.4(a). In this case the phosphor 'directly' viewed the specimen. From time to time, as the phosphor lost its ability to scintillate adequately, it would require refurbishing.

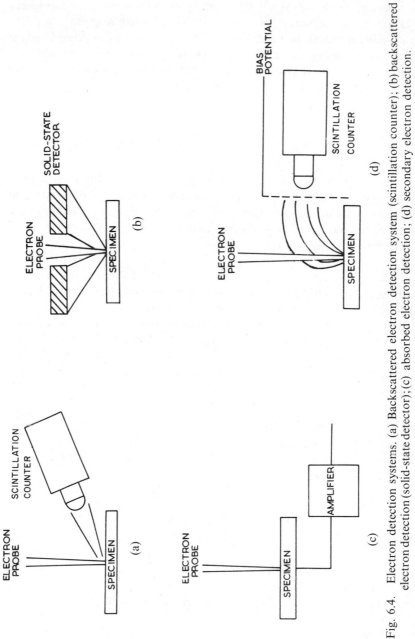

Fig. 6.4.   Electron detection systems. (a) Backscattered electron detection system (scintillation counter); (b) backscattered electron detection (solid-state detector); (c) absorbed electron detection; (d) secondary electron detection.

(b) Modern instruments tend to use solid-state detectors, frequently a solid-pair detector, viewing the specimen from different angles—Fig. 6.4(b). Again the variable pulses obtained are amplified and displayed as previously. The advantage of this method is that it allows the detected signals to be selectively combined so that either a combined atomic number (compositional) and topographical image are displayed—familiar to microanalysts; or topography is eliminated; or atomic-number effects are removed. Examples of these various types of display are shown in Fig. 6.5.

6.4.2 Absorbed Electron Detection

Provided the specimen is suitably insulated, then those electrons which are absorbed can be collected and their signal amplified—Fig. 6.4(c). The signal can be fed either to a microammeter or to a display unit. In the case of the display, the image will have the reverse contrast of the backscattered electron image—Figs. 6.5(a) and 6.5(b).

When feeding the signal to a microammeter, it acts as a very acceptable check on the stability of the electron probe, particularly if a Faraday cage is used, because in this case the total electron beam is absorbed by the cage. This check should always be used if beam instability is suspected, resulting from, for example, a contaminated column or electronic instabilities.

In certain instruments, the backscattered and absorbed electron signals are used in selective combination as another method of obtaining either a topographical or an atomic number (compositional) display. Both techniques appear to give good results, although the one described earlier is probably the better system.

6.4.3 Secondary Electron Detection

The secondary electrons are those removed from their atomic shells in the specimen being bombarded by the high-energy electrons, which by their disruption give rise to the generated characteristic X-rays of the elements present. A substantial number of these electrons escape from the specimen and a relatively simple method of detecting them is available—Fig. 6.4(d). In many respects the system is similar to the old method of detecting high-energy backscattered electrons by the use of a phosphor photomultiplier and scintillation counter.

In this case, however, the phosphor is guarded by a grid across which a $\pm$ve potential can be placed, typically in the range -200 to $+400$ V—Fig. 6.4(d).

The potential applied, therefore, attracts or repels the secondary

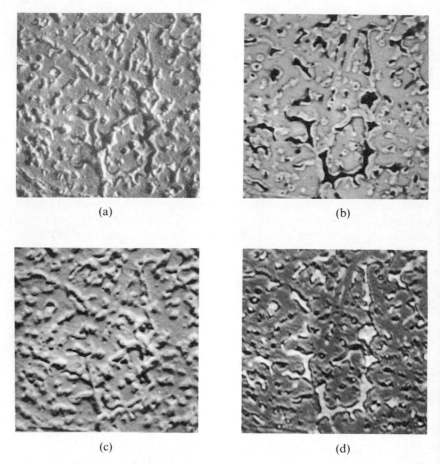

Fig. 6.5. Various types of electron imaging of alloy Cu–20 % Pb (kindly supplied by Mr G. Love of Bath University). (a) Normal backscattered electron image; (b) absorbed electron image (specimen current image); (c) topographical image (backscattered); (d) compositional electron image (backscattered-atomic number). Magnification 650 ×.

electrons, whilst the angle of the specimen controlled by the goniometer stage will affect the number of primary backscattered electrons detected. The display which is built up on the visual CRT monitor is a combination of both high- and low-energy electrons, the relative intensities of which the operator will balance in order to achieve the contrast and sample inclination necessary to optimize the details of the feature(s) of interest.

6.5   X-RAY DETECTION AND CRYSTAL SPECTROMETRY (WDX)

Microanalysis is traditionally the ability to detect characteristic X-rays by wavelength dispersion (WDX) representative of all the elements present in a feature of interest. By comparing the various intensities obtained from these elements and relating them to the X-ray intensities from appropriate standards an approximate analysis is obtained.

Chapter 7 shows how the raw analyses are corrected for such effects as characteristic and continuous fluorescence, absorption and atomic number, i.e. the ZAF correction. With the exception of the atomic number correction, both absorption and fluorescence are affected by the geometry of detection, and this is taken into account when carrying out the correction procedure. In simple terms the lower the take-off angle of the X-rays (i.e. that angle between detection and the specimen surface) the more pronounced becomes the absorption correction and the less becomes the fluorescence effect; conversely at higher take-off angles, absorption decreases and fluorescence increases.

In early instruments, the take-off angle tended to be low because of the design of the objective lens, which necessitated the X-ray path's being below the lens. Modern designs are such that almost any take-off angle is available between about 15° and 80°. In general terms, for conventional microanalysis, a higher take-off angle is to be preferred since the absorption correction is usually much greater than the fluorescence correction; hence, minimizing absorption is the more beneficial approach. For example, fluorescence may account for up to 10–15% of the correction to be applied, but absorption can be as high as 200–300%. Additionally, topographical effects, which can easily occur, for example, when preparing composites of hard and soft materials,[4,5] are minimized.

X-ray detection can be divided conveniently into three categories: (i) medium and hard X-rays, (ii) soft X-rays—in both cases by wavelength crystal or diffraction spectrometry (WDX), and (iii) energy dispersive detection techniques (EDX). In microanalysis the development in instrumentation has followed this order.

6.5.1   Medium and Hard X-ray Spectrometry

In microanalysis terms this refers to the detection of X-rays from about atomic number 11 (Na Kα) through to about atomic number 38 (Sr Lα) and for very high atomic numbers, e.g. 92 (U) when Mα detection becomes

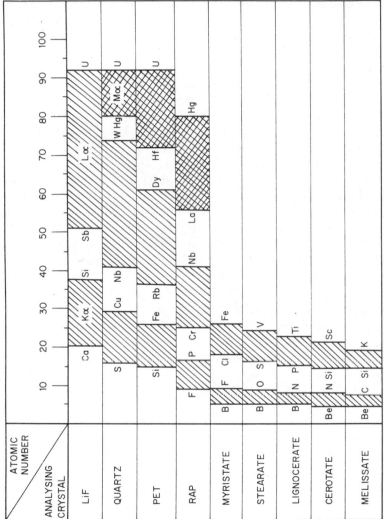

Fig. 6.6.   Typical examples of crystals and ranges of elements for which they are effective.

necessary. Figure 6.6 gives typical examples of crystals and the ranges of elements for which they are effective.

Mention should be given here of work which has been carried out on the use of X-ray diffraction gratings.[6] In this case it is said that because the crystal is not subject to the Bragg restriction of equality in the angles of incidence and reflection, it is, therefore, able to disperse all wavelengths simultaneously. This feature appears particularly attractive in measuring the emissions of low intensity and short-lived sources.[7] The grating does, however, have to be very carefully made and is usually formed by replicating an already ruled grating, as the troughs during ruling are generally of superior quality to the ridges—it is the ridges from which the diffraction occurs. Commercially, the grating method is not available, presumably because of the difficulty of manufacturing consistently good gratings.

Having designed a suitable path into the spectrometer for the X-rays, it is necessary to separate the various characteristic peaks so that individual elements can be detected and analysis carried out. There are generally three types of geometry available, and it is necessary to decide which is preferable for the general applications of the instrument.

(a) *Semi-focussing (Johann)*[8] *Geometry*
This geometry is illustrated in Fig. 6.7(a). In essence the source of X-rays (sample surface), the diffracting crystal and the detector (proportional counter) are all located on a circle (the Rowland circle), such that the diffracting crystal lattice is bent to a radius equal to the diameter of the Rowland circle. It will be seen that this does not give rise to true focussing properties. However, for many microanalytical applications this does not present a serious problem, as, for example, a mica crystal bent in this way will readily resolve the $SiK\alpha_1$ and $SiK\beta_1$ lines. In fact, whilst the fully focussing spectrometer has clear advantages for certain aspects, there is a distinct trend in many modern instruments to offer semi-focussing spectrometers. This has proved possible by the improved design of the spectrometer—giving increased X-ray detection, improved crystals and crystal types and improved proportional counters.

Semi-focussing geometry becomes necessary when the diffracting crystal cannot be ground to comply with fully focussing geometry. The particular advantages of semi-focussing geometry relate to the greater accuracy of being able to reset the correct diffracting angle (see Section 6.11) and the loss of resolution of the $K\alpha_1-\alpha_2$ peaks, for example—if these peaks are resolved it can cause analytical problems in the same way as that presented

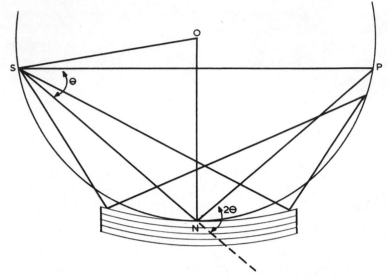

Fig. 6.7(a).   Johann focussing conditions (semi-focussing).

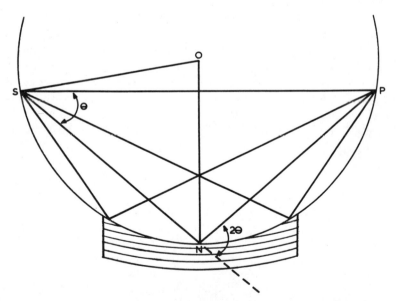

Fig. 6.7(b).   Johansson focussing conditions (fully focussing).

by overlapping peaks, e.g. SKα and MoLα. Thus analytically, and with modern design, semi-focussing spectrometry may be preferred.

(b) Fully Focussing (Johansson)[9] Geometry

The principle is the same as for semi-focussing geometry; the essential difference is that the crystal, besides being bent to a radius equivalent to the diameter of the Rowland circle, is also ground with a radius the same as the radius of the Rowland circle. This gives rise to a geometry which fully focusses the diffracted beam on to the proportional counter window—Fig. 6.7(b). For the same basic construction of spectrometer, the fully focussing system will give an improved peak:background (P:B) ratio and a better peak resolution. Both aspects are important, the former because it will improve the counting statistics, giving more rapidly and accurately available data, the latter because it will improve the resolution in the region of close or partially overlapping peaks. To obtain the full benefit, however, greater precision is necessary when 'peaking' the spectrometer.

(c) The Linear Spectrometer

The linear spectrometer can be either fully focussing or semi-focussing. The principle is illustrated in Fig. 6.8. It can be seen that a constant X-ray take-off angle is ensured. However, in this case the crystal moves along a linear path, whilst the proportional counter movement is such that the Rowland circle geometry is ensured.

The advantage of this system is that it allows for a large Rowland circle (typically 50 cm) with a reduced vacuum chamber size when compared with normal fully focussing geometry. Crystals are selected (Fig. 6.6) such that their fully focussing charactcristics are optimised about the middle of their range of wavelength detection. The slight variation from fully focussing conditions either side of this optimum does not present a problem when good-quality crystals are used in conjunction with a wide acceptance window or slits at the counter. The large Rowland circle means less bending and grinding of crystals, which itself minimizes these slight variations from fully focussing geometry. This design of spectrometer has proved very efficient, gives good resolving characteristics, is usually easily aligned and can be mounted coplanar with the electron column.

6.5.2 Soft X-ray Detection

The analysis for elements generating soft X-rays (e.g. carbon, oxygen) below atomic number ∼10, gives rise to many problems in detection, besides those met in attempting to quantify the proportions of these

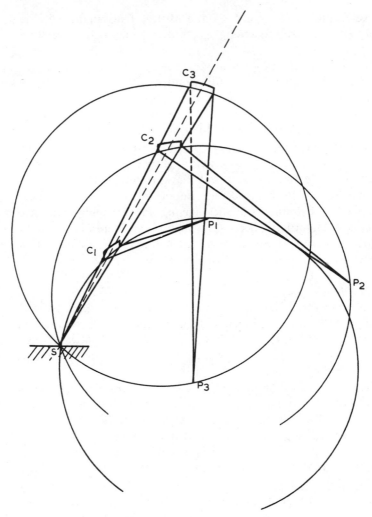

Fig. 6.8.   Schematic representation of linear spectrometer geometry.

elements present in a region of interest on the sample. Instrumentally, the problems fall into two areas:

(a) The soft X-rays, having long wavelengths, are very easily absorbed by material or gas present in the path between the source and ionization gas in the detector. Thus, the high 'take-off' angle for the X-rays becomes more important; a high vacuum in the spectrometer is essential, and as will be discussed in Section 6.6, essential windows along the path length must be sufficiently transparent.

(b) If crystal spectrometry is to be used, a suitable crystal of high 'd' spacing is necessary. The types available are shown in Fig. 6.6 and the more readily available are the stearate type (based on the soap molecule). Here, for example, an atom of Pb forms the heavy atom at the end of the molecule, and plains of these atoms are used for the diffraction process. These crystals cannot of course be ground, are very delicate and have a limited life of only a few months in service.

Electron probe microanalysis using soft X-rays has been reviewed comprehensively in a publication[10] in which it is considered that diffraction grating techniques and energy dispersive analysis (EDX) hold promise for the future.

There are other problems in the detection of soft X-rays, and a number of these are examined in the next chapter. The microanalyst will soon realize, however, that special care is needed in the selection of the standards, so that they are as similar to the specimen as possible. The threshold level, gate width (see Section 6.6) and wavelength can all vary from the optimum for, say, pure carbon, when this element is in combination with other elements, or is in a different form itself.[10]

6.6 PROPORTIONAL COUNTERS FOR X-RAY DETECTION

There are a number of different types of proportional counter, designed to meet different needs. In every case, the principle is the same. Figure 6.9(a) gives a schematic representation of a side-window counter, the type in most common usage.

The coaxial wire (anode) at a potential of 1·5–2·0 kV is surrounded by a chamber containing a gas mixture, e.g. CO_2 and argon, or methane and

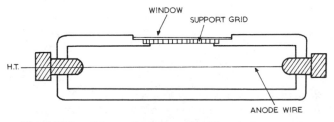

Fig. 6.9(a). Schematic design of a side-window counter.

argon, which is composed to give one gas (argon) susceptible to ionization under the bombardment of X-rays, whilst the other gas acts as both a carrier and quench gas. X-rays enter the window, causing ionization in proportion to their intensity, the ions being attracted to the anode. These ions cause small voltage drops on contact with the anode, which are recorded by the electronic system as pulses, being duly amplified and processed to give a measure of X-ray intensity (usually the output of X-ray intensity is given in counts per second—cps). These pulses can be used to modulate various display and recording systems, or, for quantitative analysis, can be integrated over a suitable time period (usually from 10–100 s) and equated to a cps value for that particular characteristic X-ray. The cleanliness of the coaxial wire is very important since the counter resolution deteriorates as the wire becomes dirty.

The process of detecting the ions at the anode begins to break down as the number of ions increases due to increased X-ray intensity. If, for example, two ions arrive simultaneously, or the second one arrives during the recovery period, then it has a reduced probability of being detected. A 'dead-time' correction has to be applied as described in the next chapter.

Of the four types of proportional counter most frequently used, three are the familiar side-window counters; the fourth type, an end-window counter, is less common.

(a) Sealed Counter (Side-window)
The sealed counter is generally used for hard or short-wavelength X-rays— Fig. 6.9(a). The window is usually of 50–100 μm beryllium; the gas, typically xenon/ethyl formate, is sealed in the counter at several atmospheres pressure and hence a continuous gas supply is not required— this would, of course, be very expensive.

(b) Flow Counter (Side-window)
The construction is the same as for the sealed counter, except for the provision of an inlet/exit system to allow a continuous flow of gas and a window of aluminized 2–3 μm mylar. The gas mixture proportions most frequently used are $2\frac{1}{2}\%$ CO_2:$97\frac{1}{2}\%$ argon; 10% methane:90% argon (P10 gas), and 75% methane:25% argon (P75 gas). The mylar window is aluminized to prevent charge collection.

(c) Light Element Flow Counter (Side-window)
This is a flow counter with the window modified to be transparent to soft X-rays. Early windows were of collodion, which easily leaked and had a short

life. Modern windows are of stretched polypropylene, again suitably metallized to avoid charging. If the window is relatively narrow, e.g. 1 mm, then no supporting grid is necessary, but for larger windows a support grid is usual and this is frequently designed to be also a collimator.

It is generally accepted that there are obvious advantages in reducing the gas density for soft X-rays[10] since the penetration is increased, thus

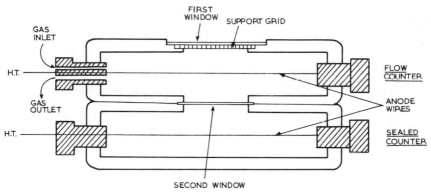

Fig. 6.9(b). Schematic arrangement of tandem counters.

allowing generated ions to escape the field distortion around the counter window and give a higher gas gain at a given voltage. It is also recognized that higher methane contents, e.g. P75, and perhaps pressures of 100–200 torr, give better counter resolution.[12]

Side-window counters are particularly useful for most microanalysis applications, because they can be mounted in tandem—Fig. 6.9(b)—i.e. one behind the other in the X-ray path. An instrument with two spectrometers may be arranged to have a standard flow counter first (with two diametrically opposite windows) followed by a sealed counter (entrance window only) in one spectrometer. In the other spectrometer a combination of a light element counter followed by a standard flow counter may be used.

(d) End-window Flow Counter

The end-window point anode counters have been investigated,[13-15] particularly for soft X-rays, using adjustable geometry (e.g. anode-to-window spacing) to optimize energy resolution. Figure 6.9(c) shows a schematic arrangement of an end-window counter. This appears to have a better resolution when compared with a side-window counter, if the

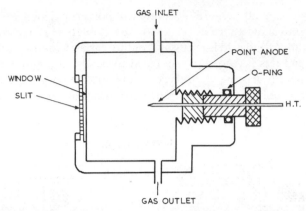

Fig. 6.9(c).   Schematic design of end-window (point-anode) counter.

geometry is optimized for a particular X-ray wavelength, e.g. window-to-anode distance. Continuous adjustment would be necessary to optimize element detection and for most operators this would cause marked inconvenience.

As there is a path along which the X-rays travel to the proportional counter, it must be appreciated that electrons can also enter the counter. These can be either primary electrons of relatively high energy which are scattered in the metallic chambers of the instrument, or are low-energy electrons emitted from the metallic surround as a result of the primary electron scattering. It is thus desirable to place slits at the entrance to the spectrometer and in front of the counter window at sufficient potential to attract these electrons, thus preventing their entering the counter and contributing to the 'background' count or noise level. By coating the inside of these slits with a light element, e.g. carbon or beryllium, the electrons are much less likely to be scattered from the surface of the slits.

The amplified pulse from the anode needs to be processed before entering the various recording and display systems. An important part of the electronics in the microanalyzer is the pulse height analyzer (PHA). An energy or pulse height display can be obtained on the pulse monitor, and by setting the correct anode voltage, the various pulses being recorded can be spread through the 5–50 V pulse range normally available. The PHA can be adjusted to any gate width (i.e. a percentage of the 45 V available), the threshold or bottom level of which can vary from a minimum of 5 V upwards. This facility allows the operator to cut out much of the background noise and X-ray continuum. Additionally, if two characteristic

X-ray peaks are at a close diffraction angle but have differing energies, giving rise to different pulse heights, then correctly set threshold and gate-width values will eliminate the unwanted peak. This happens with typically interfering first- and second-order diffractions (e.g. AlK,[1] $CrK\beta^4$); it will not differentiate $SK\alpha$ and $MoL\alpha$, for example, as these not only have similar wavelengths, but also have similar pulse height values.

6.7 X-RAY DETECTION BY ENERGY DISPERSIVE ANALYSIS (EDX)

Energy dispersive analysis for X-rays is a recent development frequently found on SEM and increasingly on conventional EPMA. For this technique a solid-state detector is used and in general the EDX system is quicker and easier to use than crystal spectrometry (WDX). However, the solid-state detector has a poorer spectral resolution and there are many situations when WDX is superior. Ideally the presence of both detection systems is desirable for a microanalytical instrument. The effective aspects of EDX are well considered by various workers.[5,16]

The basis of EDX is a Li-drifted Si detector. This semi-conductor radiation detector can have a surface area ranging up to $\sim 200 \, mm^2$, located between two electrodes across which a bias of several hundred volts can be applied, typically 500–900 V. X-rays incident on the detector create free-charge carriers by photoelectric absorption and subsequent impact ionization—the number of carriers produced being proportional to the X-ray energy and given by the ratio of the X-ray energy to the energy necessary to create a free-charge carrier (electron hole pair). The integrated current charge is subsequently amplified and electronically processed, being then passed into a multichannel analyser (MCA) where the voltage pulses are separated on a basis of amplitude and stored in memory channels appropriate to the energy level. The resulting spectrum can be displayed on a CRT as shown in Fig. 6.10 or be reproduced on a chart recorder.

The system has proved particularly useful as an addition to the SEM as it possesses adequate X-ray sensitivity at beam currents well below those used in EPMA. It will be realized that the electron beam resolution in SEM is typically 100–200 Å, but this does not mean improved X-ray resolution because of the physical restraints imposed by electron scattering and X-ray production within the sample—typically X-ray resolution from the sample is in the range 2–10 μm.

The technique has several distinct advantages, some of which have been

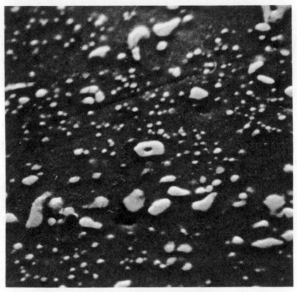

Fig. 6.10(a).   Carbides present in hispeed tool steel—M_6C type. Magnification
$1700 \times$.

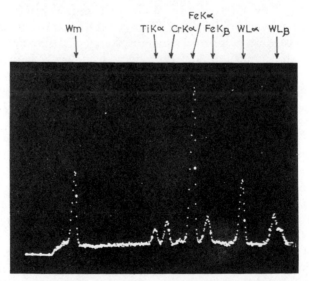

Fig. 6.10(b).   EDX display from M_6C carbide from Fig. 6.10(a) (kindly supplied
by Mr W. Johnson of Swinden Laboratories).

discussed already, and also disadvantages. Modern devices have brought the energy resolution towards 100 eV (at 6·4 keV, FeKα), and this becomes necessary to discriminate adjacent peaks. Probably the most outstanding factor is the speed with which an analysis can be performed, because the entire X-ray spectrum can be obtained easily between 30 s and a few minutes, even at moderate concentration levels. This is because the EDX system surveys the whole energy spectrum simultaneously, whilst WDX, necessitating driving the spectrometers through the total range of λ-dispersive angles, often requires $1-1\frac{1}{2}$ h for a full spectral scan. Additionally the EDX requires only one detector, compared with two or three proportional counters and a range of suitable diffracting crystals.

It can be seen, therefore, that EDX, in combination with WDX, proves a rapid and efficient system, particularly when the chemical composition of a feature is unknown. Thus a rapid EDX analysis to isolate the elements, followed by WDX at suitably selected wavelengths, will give rise to quickly produced data for quantitative analysis. Indeed, in certain places quantitative results are being obtained using EDX alone, but as will be seen in the next chapter, the theory of WDX analysis is better understood.

The EDX detector head is much smaller than a spectrometer, can be placed closer to the specimen than a diffracting crystal in the EPMA, and hence obtains a much greater solid angle of detection. The detector can be placed even closer to the specimen in the SEM because of the increased working distance between the objective pole-pieces and the specimen, and because the specimen chamber usually does not require a light-optical system.

The high sensitivity of the EDX device, allowing the use of substantially reduced beam currents, is particularly useful when examining heat-sensitive material, e.g. biological samples, and on those occasions when readily diffusable ions such as Na and K are present, where migration and indeed complete loss can occur into the vacuum system under normal WDX conditions. Obviously the higher the sensitivity, consistent with an acceptable peak:background value, the lower are the potential limits of detection; the speed of analysis also reduces instrumental problems, typically those associated with gun and electronic drift.

The absence of the focussing criterion present with WDX means that there is no serious intensity loss, even with quite large primary electron-beam deflections, or with very irregular or faceted specimens (e.g. fracture surfaces). The EDX system is also mechanically very simple with no moving parts and does not require the extreme precision encountered with the WDX spectrometer and all the features these require.

The EDX system is not without its disadvantages, however. In particular the Si detector and FET transistor must be kept at cryogenic temperatures for normal operating conditions when a bias is applied to the detector. The cryostat is rather bulky, requires refilling with liquid nitrogen, but a 5-litre one will last 4 to 5 days. Because of the low temperature of the detector surface, it must be maintained in an evacuated chamber which is isolated from the main instrument vacuum. It is usual to use a Be window for this, although more recently metalized polypropylene is being favoured because of its better transparency to soft X-rays. In these circumstances, a support grid and isolation valve for window failure would be necessary. The isolating window is needed to prevent condensation of water and oil vapours on the detector head and because the detector is light sensitive. The window also acts to some extent as a backscattered electron absorber and a vacuum seal between the EDX and the column. If they can be suitably arranged, slits in front of the detector, Be coated on their inside surfaces, and at a positive bias potential will minimize the number of electrons reaching the detector.

Whilst the EDX system has so far been described as a desirable facility to run in tandem with WDX in microanalysis, or on its own in SEM, clearly if it can be used for quantitative analysis,[17] then even greater benefit can be obtained—this aspect is examined in the next chapter. It suffices to record here that the two main problems met are: (i) the deconvolution of the energy spectrum, which requires computer application for anything other than a binary analysis because of the generally poor spectral resolution and P:B ratios; (ii) the accurate determination of the background intensity, which can constitute a significant portion of the peak and which does not vary continuously with energy.

6.8   DISPLAY AND RECORDING SYSTEMS

There are a variety of display and recording systems available commercially, some of which are more relevant to EPMA whilst others have greater application in SEM.

6.8.1 Display Systems
Most microanalysers have two display CRTs with square-faced presentation. This is principally to allow simultaneous electron and X-ray images to be displayed. The CRTs are of medium to low phosphor persistence to allow both easy viewing, and also rapid reimaging, after

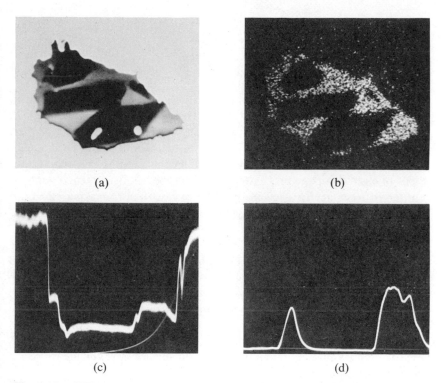

Fig. 6.11.   EPMA display modes (MnO–Al$_2$O$_3$–SiO$_2$ inclusion in steel). (a) Electron image (backscattered); (b) X-ray image (SiKα); (c) electron monitor (backscattered); (d) X-ray line scan (SiKα).

specimen movement. The following facilities are necessary on the CRT display for EPMA.

(a) Interchangeable image between the two CRTs; this is particularly useful for superimposing an X-ray image on a fading electron image to locate areas of element concentration. It is also preferable to record photographically features from the same CRT in order to standardize conditions, and to use the CRT with the best resolution.

(b) The operator should be able to display any electron or X-ray spatial image of his choice (Figs. 6.11(a) and (b)) and be able to have a continually variable $X–Y$ raster size and position variation to concentrate examination on a particular feature without altering magnification. Magnification should be continually variable between 10 and 10 000 times approximately.

(c) The facility of monitoring the backscattered electron intensity along a line is essential for focussing (Fig. 6.11(c)). Electronic line scanning of the electron beam for X-ray intensity allows the variation in element concentration to be rapidly appraised, leading to the identification of features of interest (Fig. 6.11(d)).

(d) Mention has already been made of the various forms of electron image which can be obtained and clearly these require display facilities, i.e. backscattered, absorbed and secondary electron images and their combinations to give topographical and atomic number images. Additionally, although probably more relevant to SEM, are the various types of electron image processing. In both types of instrument a contrast limiter is necessary to prevent 'bright' features from masking effects or burning the CRT phosphor. Typical of the image processing facilities available are:

(i) *Expanded contrast* (*electron images*). This is particularly useful where, for example, there are regions of marked contrast difference such that conditions for observing one region mask all the features in another. This technique allows the full black–white contrast range of the CRT to be used for each region to accentuate the small signal changes, at the same time equating each region to a mean intensity level across the whole CRT, e.g. the full detail would be shown from a bright surface area and a dark recess of re-entrant hole.

(ii) *Contours* (*electron images*). In this case a bright contour line is generated against a black background for preset variations in signal level. It allows small changes in signal level to be much more positively displayed than a small change in image brightness.

(iii) *Contours* (*X-ray images*). Whilst not commonly available, contouring of X-ray intensities on the CRT display is not difficult, and can prove a useful display facility. The fact that the corrections for a particular element will vary with composition across the area being examined limits the usefulness of the system.

(iv) *Grey level* (*electron images*). This display will emphasize the structure, removing the confusion of fine detail, i.e. within a series of predetermined signal bands, that particular signal will be presented as a uniform grey. The technique is used particularly to emphasize atomic number contrast in multiphase systems.

A facility which can be of particular use to the microanalyst is the storage CRT. This has a storage grid within the CRT, such that the input built up over a series of scans can be stored and displayed as a static image of the integrated scans. This method is particularly useful for displaying low concentration profiles or minor segregation effects as indicated by the spatial X-ray image. This is an alternative to photography for location of regions of interest and is much more rapid. This form of display can be turned on and off as required, has a persistence of several hours, and the stored signal is readily cancelled. A particularly useful application is the ability to build up an X-ray spatial image on an already displayed electron image.

It is also possible to interface both EPMA and SEM with a TV scanning mode and present the image this way. Generally, some resolution is lost, but the method has several advantages, including reduced eye strain and the ease of viewing the image by a number of people.

6.8.2 Recording Systems

Photography has been mentioned, and it is considered that good and easily handled photographic facilities should be available, ideally including both 35 mm or roll film and polaroid type.

Besides the ability to photograph the visual displays, it is also necessary to record the data which will be used for analysis. All EPMAs have scalers available for each WDX and EDX channel with timers integrating over steps typically 10, 30, 100 s, up to many minutes. Additionally, many instruments have the facility to record a preset number of counts, e.g. 10 000, and record the time taken. This can be particularly useful when optimising the counting time for analysis.[18]

The data available from the scalers can be output in several useful forms, either numerically by an interfaced typewriter, or on to cards, punched tape or magnetic tape for computer processing.

When carrying out segregation profiles, it may be preferable to chart-record the results continually from each spectrometer channel. This can be done in several ways: (i) by a multi-offset pen system; (ii) by a dotting system alternating between the various channels; or (iii) by using an $X-Y$ recorder. Each of these techniques has its advantages and disadvantages. The dotting system, whilst not producing a continuous trace, does have the advantage of not offsetting the scans in relation to each other on the chart.

It is also necessary to be able to monitor the beam/specimen current in order to ensure instrument stability and quickly isolate any drift which will affect the analyses. Additionally, variable range ratemeters are essential in order to select the most appropriate range (full scale deflection, FSD) for a particular analysis and to peak the spectrometers.

6.9   VACUUM SYSTEMS

Often the vacuum system tends to be an accepted part of the instrument. However, it is a very important aspect and needs good maintenance in order to ensure good operating conditions, particularly in relation to filament life, electron beam resolution and specimen contamination, which may seriously affect analysis reliability.

In modern instruments, an integrated vacuum system is generally available with automatic control and fail-safe devices. This is particularly important because back-streaming of vacuum pump oils, when air is admitted incorrectly, can cause considerable damage and time loss in cleaning out the whole vacuum system.

Generally, the better the vacuum system the more reliable will be the instrument and the examinations carried out. It is therefore important to maintain the system to the manufacturer's specifications—or better than these if possible. Section 6.2 described the typical orders of vacuum necessary.

When examining an instrument's vacuum system for suitability, several general points should always be borne in mind:

1. A minimum number of vacuum seals reduces the chances of a leak occurring.
2. The diffusion and backing pumps should be of sufficient capacity to cope adequately with the vacuum system and have capacity to spare.
3. Spectrometers and column should have separate pumping lines which can be integrated when operating conditions have been achieved.
4. Pumping efficiency depends mainly on the bore and straightness of the pumping lines assuming the pumps are operating correctly.
5. Valves should be incorporated to isolate different parts of the system should a leak occur, or when the gun, spectrometers or specimen chamber need to be brought to atmospheric pressure.
6. The vacuum system controls should be fully integrated and automatic, arranged on a display panel to show the state of the vacuum system, and be located in close proximity to the vacuum gauge indicators.

High-vacuum technology is a specialized subject in its own right, and hence only the broader features have been discussed. However, it should be mentioned that in addition to oil diffusion pumps, ion diffusion pumps are available, capable of giving a higher ultimate vacuum.

6.10   AUTOMATION OF EPMA

Automation is clearly more applicable to EPMA than SEM because many operations can become very repetitive. Various stages of automation have been achieved in modern instruments, but the stage has been reached where full automation is the next major development and shortly instruments should be available that largely achieve this objective. Any microanalyst will realize that there are a multitude of things that can go wrong during an analysis, and therefore any proposed system must be able to carry out these operations. The computer use could be further extended to process the necessary corrections on the raw data, presenting true concentrations as the output. Computerizing to this extent would remove much of the routine nature of the instrument, but nevertheless the operator would still need to be highly trained to be able to recognize, understand and correct anomalies that occur.

There is no doubt that having decided upon the examination required, and put the instrument into a 'go' situation, computerization could carry out all the necessary steps needed to obtain the results. This would include searching the specimen for features of interest, based on X-ray output, peaking of spectrometers and compensating for instrument drift, in addition to the other processes necessary for analyses to be obtained. The keyboard of the typewriter output would most probably act as the input command unit for the computer, although a more sophisticated punched tape or card input may be appropriate so that a series of command programs could be utilized for various forms of analysis and search.

6.11   INSTRUMENTAL ACCURACY

It will be obvious when considering the various processes that occur between the generation of the high-energy electron beam and the final analyses that there are many parts of the instrumentation that are subject to error. Work has been carried out[11,19,20] that has examined, under practical operating conditions, the typical errors that may be expected, based on results using metallic or oxidic specimens. For organic materials the errors associated with the specimen will be accentuated.

It was found that the specimen finish could be a significant problem for long wavelengths and low take-off angles, but below $0.25\,\mu$m finish, in the absence of differential polishing, no significant variations were found.

The location of the specimen in relation to both height and tilt shows significant errors, the former because it takes the surface off the Rowland circle, and the latter because it will affect X-ray intensities in multicomponent systems—this is due to the absorption path within the sample varying as the tilt occurs in relation to the normal take-off angle.

Contamination of the specimen markedly alters X-ray intensities, even, for example, NiKα. The results obtained showed marked differences between the various instruments, ranging from an intensity loss of $1\frac{1}{2}$–8% after 15 min, and 3–14% after 30 min for NiKα. This sort of loss is very important and suggests much greater attention is required to maintain good vacuum systems than is perhaps normally practised. The use of the cold finger will of course decrease this contamination rate, and it is essential for soft X-ray work.

During a complete analysis the spectrometer is moved many times and needs repositioning with high accuracy, particularly if it is of the fully focussing type. If the instrument is automated, then a peak-seeking routine becomes necessary. Manufacturers generally claim repeatability between ± 5 and ± 60 s of arc. Specimen height and Bragg peak sharpness will also have an effect. In testing a group of spectrometers[11] relocation gave an average error of 0.46% for Ni using a LiF crystal. Temperature changes in the crystal will affect the Bragg angle as well as the mechanical linkages in the spectrometer, e.g. the coefficient of expansion of LiF is $31 \times 10^{-6}\,°C^{-1}$ and for quartz is $14 \times 10^{-6}\,°C^{-1}$ perpendicular to the Z axis. Thus it can be shown that a 1 °C temperature change corresponds to 6 s of arc for LiF at $2\theta = 86°$ and typically 50% of peak intensity would be lost by a 2-min movement off peak in this part of the spectrum. It has already been pointed out that EDX would overcome these problems—but the longer-term stability of EDX has yet to be proved.

Short- and long-term stability of the electron probe is important. An estimate of the short/medium-term stability (5 min) over a number of instruments was determined at 0.3% coefficient of variation. This value compares with manufacturing claims of 0.02% for 15 min to 0.4% per hour. Therefore questions arise as to the condition of the equipment or the environment in which it exists. Most microanalysis is carried out using a ratio technique to obtain analyses, and this will minimize stability errors in the system, including of course the longer-term stability of the gas proportional counters.

6.12   CONCLUSIONS

In compiling this chapter on instrumentation, it has been possible only to outline the various features involved, introducing some detail where appropriate, but mainly indicating desirable aspects. The improvements in design, reliability, ease and range of operation have advanced considerably since the first commercial microanalysers appeared in the late 1950s. Many other electron-optical techniques have emerged, including the scanning electron microscope which is compatible with microanalysis, and the combination of both techniques in one instrument is common and for many situations very desirable. Advances in EDX hold great promise, particularly if accurate quantitative analysis becomes possible, but it will need to be at least comparable with WDX. Highly automated instruments are now emerging as commercial propositions, with computer control available to the extent of correcting the raw data to give true elemental concentrations.

The choice of which instrument to purchase is a difficult one. The decision must be based on many factors including, principally, suitability for use in relation to the initial cost, and reliability, not forgetting the servicing facilities offered. As instruments become more sophisticated, it makes it increasingly necessary to rely on these manufacturers' services. Thus a careful survey is an essential prerequisite to purchasing, and experiments should be conducted to adequately test performance, at least to the level required during the useful life of the instrument.

REFERENCES AND BIBLIOGRAPHY

1. H. E. Bishop, *Advances in Analysis of Microstructural Features by Electron Beam Techniques*, Met. Soc. Pub. of Conf. Proc., pp. 1–18, 1974.
2. D. Joy, *ibid.*, pp. 20–40.
3. L. H. Veneklassen, *Optik*, 1972, **36**(4), 410–33.
4. W. J. M. Salter, *A Manual of Quantitative Microanalysis*. Structural Publications, London, 1970.
5. D. R. Beaman and J. A. Isasi, *Electron Beam Microanalysis*, ASTM–STP 506, 1972.
6. A. Franks and K. Lindsay, *The Electron Microprobe*, pp. 83–92. Wiley, New York, 1966.
7. M. Stedman, *Advances in Analysis of Microstructural Features by Electron Beam Techniques*, Met. Soc. Pub. of Conf. Proc., p. 119, 1974.
8. H. H. Johann, *Zeit. für Physik*, 1931, **69**, 185.
9. B. V. Johansson, *Micron*, 1973, **4**, 121–35.

10. V. D. Scott, *Advances in Analysis of Microstructural Features by Electron Beam Techniques*, Met. Soc. Pub. of Conf. Proc., pp. 141–92, 1974.
11. W. J. M. Salter and W. Johnson, *Micron*, 1970, **2,** 157–86.
12. A. A. MacFarlane, *Micron*, 1972, **3,** 506–25.
13. P. Duncumb and D. A. Melford, *Metallurgia*, 1960, **61,** 205–12.
14. E. Mathieson and P. W. Sanford, *J. Sci. Inst.*, 1963, **40,** 446–9.
15. G. V. T. Ranzetta and V. D. Scott, *J. Sci. Inst.*, 1967, **44,** 987.
16. M. H. Jacobs, *Advances in Analysis of Microstructural Features by Electron Beam Techniques*, Met. Soc. Pub. of Conf. Proc., pp. 80–118, 1974.
17. S. J. B. Reed and N. G. Ware, *X-ray Spectrometry*, 1973, **2,** 69–74.
18. N. Swindells, *Advances in Analysis of Microstructural Features by Electron Beam Techniques*, Met. Soc. Pub. of Conf. Proc., pp. 214–17, 1974.
19. W. Johnson, *ibid.*, ref. 18, pp. 231–54.
20. W. Johnson and W. J. M. Salter, *Micron*, 1973, **4,** 87–97.

7

Quantitative Microanalysis

C. W. HAWORTH
University of Sheffield, UK

7.1 INTRODUCTION

Quantitative analysis depends on measurement of the intensity of characteristic X-rays. Whilst the principles of X-ray production are easily understood, the relationship between intensity, I_A, and concentration of the corresponding element, C_A, is complex. Various approximations must be used in establishing this relationship, and there are many alternative equations and methods of computing C_A in the literature. A comprehensive review is available[1] giving some of the rival derivations, but for practical microanalysis it is more important that one reliable procedure is understood, together with the errors and approximations inherent in the calculation, so that operating conditions and procedure can be chosen to minimize the consequences of such uncertainties.

Most microanalysis is performed by comparison between the intensity of characteristic radiation from the specimen I_A and the intensity from a pure element or well-defined compound standard $I(A)$. The ratio

$$K_A^m = I_A^m / I^m(A)$$

where the superscript indicates the intensity of radiation measured by the detection system, is then the starting point for the analysis. The relationship with weight fraction is then derived either by calculating the number of characteristic X-ray quanta produced per electron, which is perhaps the most fundamental approach, or more commonly an approximately proportional relationship is assumed to exist between K_A^m and the weight fraction of element A, C_A, and this is written in the form

$$K_A^m = (ZAF)C_A$$

157

where the factors Z (atomic number correction), A (absorption correction) and F (fluorescence correction) arise from the physical basis of the different corrections. A similar equation, with different values of Z, A and F holds for the other elements B, C, etc. present in the specimen. Z, A and F are each functions of composition, and so whilst K_A^m can be directly calculated for a given value of C_A, C_B, C_C ... the reverse process (and calculating C_A from a measured value of K_A^m is the usual analytical situation) requires an iterative procedure. This method of microanalysis starting with K_A^m can give a result which is accurate to within 5% of content and is usually better than this.

An alternative approach to microanalysis, which is particularly suitable for analysis using an energy dispersive spectrometer (EDX) where all wavelengths are measured simultaneously, depends on measurement of the ratios I_A: I_B: I_C ... for all the different elements present in the specimen. The total concentration is then assumed to sum to 100%, but it must be stressed that this ratio of intensities cannot be taken (even approximately) as equal to the ratio of concentrations. The technique assumes that the measured concentrations sum to a known value (usually 100%), and requires calibration of the system for excitation and detection efficiency, in addition to the ZAF calculation, and the overall accuracy is somewhat worse than the procedure based on the measurement of the ratio K for each element. A variant of this technique must also be employed for the microanalysis of a thin foil specimen, as used in the transmission electron microscope, where the thickness of the specimen is a variable.

If analysis with the best possible accuracy is required, then the use of a homogeneous standard is indicated, the standard being chosen to be as similar as possible to the specimen being analysed. A method suitable for processing the data in this case is based upon the work of Ziebold and Ogilvie[2] and can give an accuracy of about 1% of content, but this technique depends upon the production of standard specimens which are homogeneous on a micro scale and on the chemical analysis of the standard.

Whichever method is used for computing C_A, the analysis ultimately depends on the measurement of an X-ray intensity. This must be corrected for the background radiation, for any non-linear response of the measuring system (such as dead-time losses) and for any other source of error. Any error in the intensity measurement will obviously lead to an error in the analysis and a good understanding of the method of operation of the X-ray detection system is essential. Appreciation of the importance of specimen preparation and possible changes brought about in the specimen by electron bombardment is equally important. These experimental aspects of electron probe microanalysis have been discussed in detail by Sweatman and Long[3] and by Beaman and Isasi[4] and for the microanalyst these articles will repay detailed study. The more important aspects of experimental techniques are given in Sections 6.11 and 7.9.

7.2   PRINCIPLES

Electrons are slowed down and lose their energy in the specimen, and characteristic X-rays are produced by several processes:

(a) *Primary X-rays* are produced as a result of the direct ionization of atoms by the *incident electrons*. This is the process responsible for most of the emitted radiation. It can occur provided the energy of the electron $E > E_c$, the energy of the appropriate ionization level of the atom. The corresponding X-ray wavelengths are emitted when the excited atom decays to the ground state.

The number of quanta produced per electron depends on:

(i)   the ionization cross-section $Q(E)$ for the atom being ionized; this is a function of the energy of the electron, as shown in Fig. 7.1;

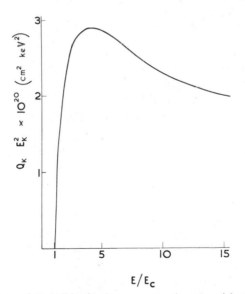

Fig. 7.1.   Variation of K shell ionization cross-section Q_K with electron energy E (after Green and Cosslett).[18]

C. W. HAWORTH

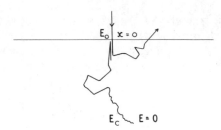

Fig. 7.2. Typical electron paths in a solid specimen. Electrons enter with energy E_0, and Q_K varies with E along the electron path x. One backscattered electron and one path terminating in the specimen are shown.

(ii) the rate at which the electron loses energy along its path length ρx in the specimen of density ρ; this is termed the stopping power of the specimen,

$$S = -\frac{dE}{d(\rho x)}$$

and

(iii) the loss of ionization, because a fraction η of the electrons are backscattered from the specimen. The magnitude of this effect depends on the energy of the backscattered electrons since only those backscattered with $E > E_c$ result in loss of ionization. The factor R by which backscattering reduces ionization is then given by $R = 1 - \eta$.

The number of ionizations produced per electron is thus calculated by integrating along the electron path for Q and S, (Fig. 7.2) and then including a multiplying factor R. If there are dn_A ionizations of A atoms in an element of electron path dx for which the electron energy falls from E to $E-dE$, then

$$dn_A = (\text{number of A atoms per unit volume}) \times Q_A(E)\,dx$$

$$= \frac{N\rho C_A}{A_A} Q_A(E)\,dx$$

where N is Avogadro's number, A_A is the atomic weight of element A which is present at weight fraction C_A in a specimen of density ρ. Integration along the electron path is best carried out by changing the variable to E when

$$n_A = \frac{NC_A}{A_A} \int_{E_0}^{E_c} \frac{Q(E)}{S(E)}\,dE$$

where $S(E)$ is defined above.

Making allowance for the backscatter effect R, for the fluorescence efficiency ω_c (the fraction of ionized atoms that emit X-rays) and for the quantum weighting ξ (the fraction of the emitted characteristic X-ray intensity that is in the line being measured) we have the intensity produced

$$I_A = \frac{NC_A}{A_A} R\omega_c\xi \int_{E_0}^{E_c} \frac{Q(E)}{S(E)} dE$$

In this expression Q is the ionization cross-section for the radiation being measured. S and R depend upon both the radiation being measured and the atomic number of the specimen, and consequently this correction is referred to as the *atomic number* or Z correction.

The term Z in the ZAF correction can therefore be written

$$Z = \frac{R_A \int_{E_0}^{E_c} \dfrac{Q_A(E)}{S_A(E)} dE}{R_{Std} \int_{E_0}^{E_c} \dfrac{Q_{Std}(E)}{S_{Std}(E)} dE}$$

where the numerator relates to the production of characteristic X-rays from element A in the specimen and the denominator to the production of the same wavelength radiation in the standard.

The various procedures that are used to calculate the value of I_A differ in the approximations used for the parameters Q, S and R, and in the way in which the integration over E is performed.

(b) The radiation produced in the specimen suffers absorption as it emerges, and consequently the radiation that is measured, I_A^m, is less than that which is produced, I_A. The allowance for this absorption (which occurs in both the specimen and the standard) is termed the *absorption correction*, A, and is made by calculating a function $f(\chi)$ where $\chi = \mu \operatorname{cosec} \theta$, μ being the mass absorption coefficient of the sample for the radiation being measured and θ being the X-ray take-off angle from the specimen surface.

From a specimen, if $\phi_A(\rho z)\,d(\rho z)$ is the intensity generated in a layer of mass thickness $d(\rho z)$ at a mass depth of ρz (see Fig. 7.3), then this radiation must pass through a length $\rho z \operatorname{cosec} \theta$ and consequently suffer absorption and be reduced according to Beer's law by a factor

$$\exp - (\mu\rho z \operatorname{cosec} \theta) = \exp(-\chi\rho z)$$

Therefore

$$I_A^m = \varepsilon C_A \int_0^\infty \phi_A(\rho z) \exp(-\chi\rho z)\, d(\rho z)$$

where ε is a factor to allow for the efficiency of the detection equipment. The integral is usually termed $F(\chi)$ and thus the measured ratio

$$K_A = \frac{I_A^m}{I^m(A)} = C_A \frac{F(\chi_A)}{F[\chi(A)]}$$

the numerator being calculated for the specimen, the denominator for the standard.

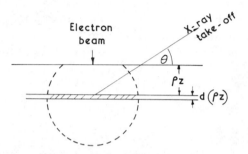

Fig. 7.3. Intensity generation function $\phi(\rho z)$ is expressed in terms of mass depth in the specimen ρz, and is used in calculating the effect of X-ray absorption in the specimen. Dotted line indicates volume in which electrons cause direct generation of X-rays.

In principle the function ϕ could be derived from the theory relating to X-ray production (Fig. 7.2). However, in practice it is not possible to derive an accurate value for $\phi(\rho z)$, and hence $F(\chi)$, from theory alone, and most correction procedures are based on empirical data. Experimental measurement of $\phi(\rho z)$ (Fig. 7.4) involves the use of thin evaporated layers of a tracer element covered by evaporated layers of specimen of various thicknesses. In principle the terms $\phi(\rho z)$ measured with electron penetration z, and $Q(E)/S(E)$ measured along the electron path x, are representing the same quantity. The two separate functions are necessary because of the inadequacies of the approximate methods used in their evaluation. $\phi(\rho z)$ is intended to be a good representation of variation of ionization with depth, whilst S describes the variation of ionization with atomic number. Thus $\phi(\rho z)$ is used to calculate the function

$$f(\chi) = \frac{F(\chi)}{F(0)} = \frac{\text{intensity emitted}}{\text{intensity produced}}$$

where the use of F(0) takes care of errors in $\phi(\rho z)$ and the function $f(\chi)$

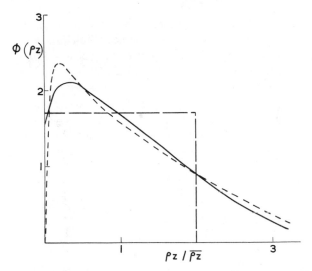

Fig. 7.4.   Intensity generated as a function of mass depth in the specimen. Full line—experimental ionization function; dotted line—Philibert[12] approximation, $\phi(0) = 0$; dashed line—Bishop,[13] the 'rectangular' approximation.

makes correction for absorption alone. The 'A' term in the ZAF correction is then given by

$$A = \frac{f_A(\chi)}{f_{Std}(\chi)}$$

where in the numerator χ is calculated for absorption of radiation of element A in the specimen and f_A is the absorption function calculated for the specimen. In the denominator χ and f_{Std} are calculated for the standard.

(c) *Fluorescent X-rays* arise when the initial ionization is brought about by absorption of an X-ray quantum rather than an electron. Two possible processes can occur:

(i)   Ionization by the white or *continuous X-ray spectrum* which is produced as the incident electrons lose energy by collisions with atoms. That part of the white spectrum with $E > E_c$ will always provide such a contribution, I_c^f, but it is only believed to be significant when higher atomic number elements ($Z > 35$) are present in the specimen. It is never greater than a few per cent and is often ignored as far as electron probe microanalysis is concerned.

(ii)   Ionization by absorption of *characteristic* X-rays coming from another element in the specimen. This will only happen if the energy of the radiation (from another element such as B) is greater than the E_c value for the radiation that is being measured (from element A), and in this case there is a contribution I_A^f to the emitted intensity. The effect is most significant when the energy of the radiation causing fluorescence is *just* greater than E_c and when a small concentration of A is being measured, in which case 20% of the emitted characteristic radiation can arise from this process. The correction is obtained by computing the sum of all these contributions,

$$\gamma_{cA} = \frac{I_{cA}^f}{I_A^m}$$

being due to the continuous spectrum, and

$$\gamma_{jA} = \frac{I_{jA}^f}{I_A^m}$$

being due to ionization caused by emission of characteristic X-rays from element j.

The fluorescence factor in the correction is then

$$F = (1 + \sum \gamma)$$

and the entire correction is written

$$K_A^m = ZAFC_A$$

7.3   *ZAF* CORRECTION PROCEDURE

7.3.1 Z, the Atomic Number Correction

$$Z_A = \frac{R_A \int_{E_0}^{E_c} \frac{Q_A(E)}{S_A(E)} \, dE}{R_{Std} \int_{E_0}^{E_c} \frac{Q_{Std}(E)}{S_{Std}(E)} \, dE}$$

has to be computed for each element that is measured (e.g. A above) and for each element that is present.

Values for R, *the backscatter loss factor*, lie in the range $0 \cdot 5$–$1 \cdot 0$ and

$R \to 1$ when: (a) the *specimen* has a low atomic number and consequently there is little backscattering, or (b) the *accelerating voltage* E_0 is only just greater than the ionization potential E_c, so that any interaction with the specimen results in an electron energy $E < E_c$ and consequently no loss of ionization results from such backscattered electrons.

Values for R have been derived from experimental measurements of electron backscattering by Bishop[5] and have been fitted by a polynomial (Springer,[6] see Table 7.1). Each value depends upon Z, the atomic number of the element present and $U_0 = E_0/E_c$, the overvoltage ratio for the particular line being measured. For the K lines the critical voltage is E_K; for the $L\alpha$ line it is the energy of the appropriate L_{III} absorption edge, and for the $M\alpha$ line it is the M_V edge. These data are in Table 7.2. An R value for the specimen when measuring element A is calculated from the relation R_A (specimen) $= R_A(A)C_A + R_A(B)C_B + \cdots$ where $R_A(B)$ is the backscatter factor for the U value corresponding to the A radiation being measured and for the atomic number of element B. This method of averaging is adequate at values of E_0 above 20 keV, but the work of Bishop has indicated discrepancies when heavier elements are involved and lower voltages are used.

The number of photons per electron is calculated by evaluating the integral

$$\int_{E_0}^{E_c} \frac{Q}{S} \, dE$$

for the ionization cross-section Q and the stopping power $S = -dE/d(\rho x)$. We can use expressions proposed theoretically by Bethe *et al.*[7] and slightly modified by Nelms[8]

$$Q = \frac{0 \cdot 76 \pi e^4 n}{E_c^2} \times \frac{\ln U}{U}$$

$$S = 2\pi e^4 N \rho \frac{Z}{A} \times \frac{1}{E} \ln \left(\frac{1 \cdot 166 E}{J} \right)$$

where n is the number of electrons in the ionization shell of the atom, e is the electronic charge, N is Avogadro's number and J is the mean ionization potential for the atom. There is some doubt about the best expression for J. The approximation $J = 11 \cdot 5Z$ has been used, but Berger and Seltzer[9] fitted the expression

$$J = 9 \cdot 76Z + 58 \cdot 5Z^{-0.19}$$

TABLE 7.1
Values of R as a Function of Atomic Number Z and Overvoltage U, computed from Springer's Polynomial[6]

Z \ 1/U	0·01	0·1	0·2	0·3	0·4	0·5	0·6	0·7	0·8	0·9	1·0
1	0·994	0·996	0·997	0·998	0·998	0·999	1·000	1·000	1·000	1·000	1·000
2	0·988	0·991	0·993	0·995	0·996	0·998	0·999	1·000	1·000	1·000	1·000
3	0·982	0·986	0·989	0·992	0·994	0·996	0·998	0·999	1·000	1·000	1·000
4	0·975	0·981	0·985	0·988	0·991	0·994	0·996	0·998	0·999	1·000	1·000
5	0·968	0·975	0·980	0·984	0·988	0·991	0·994	0·997	0·999	1·000	1·000
6	0·961	0·969	0·975	0·980	0·984	0·989	0·992	0·996	0·998	0·999	1·000
7	0·954	0·963	0·970	0·975	0·981	0·986	0·990	0·994	0·997	0·999	1·000
8	0·947	0·956	0·964	0·971	0·977	0·982	0·988	0·992	0·996	0·999	1·000
9	0·940	0·950	0·959	0·966	0·972	0·979	0·985	0·991	0·995	0·998	1·000
10	0·932	0·943	0·953	0·961	0·968	0·975	0·982	0·989	0·994	0·998	1·000
11	0·925	0·936	0·947	0·955	0·963	0·972	0·979	0·986	0·992	0·997	1·000
12	0·917	0·930	0·940	0·950	0·959	0·968	0·976	0·984	0·991	0·996	1·000
13	0·909	0·923	0·934	0·944	0·954	0·964	0·973	0·982	0·990	0·996	1·000
14	0·902	0·916	0·928	0·939	0·949	0·959	0·970	0·979	0·988	0·995	1·000
15	0·894	0·909	0·921	0·933	0·944	0·955	0·966	0·977	0·986	0·994	1·000
16	0·886	0·902	0·915	0·927	0·939	0·951	0·963	0·974	0·985	0·994	1·000
17	0·879	0·894	0·909	0·921	0·934	0·946	0·959	0·972	0·983	0·993	1·000
18	0·871	0·887	0·902	0·915	0·928	0·942	0·956	0·969	0·981	0·992	1·000
19	0·864	0·880	0·896	0·909	0·923	0·937	0·952	0·966	0·979	0·991	1·000
20	0·856	0·873	0·889	0·904	0·918	0·933	0·948	0·963	0·978	0·990	1·000
21	0·849	0·867	0·883	0·898	0·913	0·928	0·944	0·961	0·976	0·989	1·000
22	0·842	0·860	0·876	0·892	0·907	0·924	0·941	0·958	0·974	0·988	1·000
23	0·835	0·853	0·870	0·886	0·902	0·919	0·937	0·955	0·972	0·987	1·000
24	0·828	0·846	0·864	0·880	0·897	0·915	0·933	0·952	0·970	0·986	1·000
25	0·821	0·840	0·858	0·875	0·892	0·910	0·929	0·949	0·968	0·985	1·000
26	0·814	0·833	0·852	0·869	0·887	0·906	0·926	0·946	0·966	0·984	1·000
27	0·807	0·827	0·846	0·864	0·882	0·901	0·922	0·943	0·964	0·984	1·000
28	0·801	0·821	0·840	0·858	0·877	0·897	0·918	0·940	0·962	0·983	1·000
29	0·794	0·815	0·835	0·853	0·872	0·893	0·915	0·937	0·960	0·982	1·000
30	0·788	0·809	0·829	0·848	0·868	0·889	0·911	0·935	0·958	0·980	1·000
31	0·782	0·803	0·824	0·843	0·863	0·884	0·908	0·932	0·956	0·979	1·000
32	0·776	0·798	0·818	0·838	0·858	0·880	0·904	0·929	0·954	0·978	1·000
33	0·770	0·792	0·813	0·833	0·854	0·876	0·901	0·926	0·952	0·977	1·000
34	0·764	0·787	0·808	0·828	0·849	0·872	0·897	0·923	0·950	0·976	1·000
35	0·759	0·782	0·803	0·824	0·845	0·868	0·894	0·921	0·948	0·975	1·000
36	0·754	0·776	0·798	0·819	0·841	0·865	0·890	0·918	0·946	0·974	1·000
37	0·748	0·771	0·794	0·815	0·837	0·861	0·887	0·915	0·945	0·973	1·000
38	0·743	0·767	0·789	0·810	0·833	0·857	0·884	0·913	0·943	0·972	1·000
39	0·738	0·762	0·785	0·806	0·829	0·854	0·881	0·910	0·941	0·971	1·000
40	0·734	0·758	0·780	0·802	0·825	0·850	0·878	0·908	0·939	0·970	1·000
41	0·729	0·753	0·776	0·798	0·821	0·847	0·875	0·905	0·937	0·969	1·000
42	0·725	0·749	0·772	0·794	0·818	0·843	0·872	0·903	0·935	0·968	1·000
43	0·720	0·745	0·768	0·791	0·814	0·840	0·869	0·900	0·933	0·967	1·000
44	0·716	0·741	0·764	0·787	0·811	0·837	0·866	0·898	0·932	0·966	1·000

TABLE 7.1—*contd.*

Z \ 1/U	0·01	0·1	0·2	0·3	0·4	0·5	0·6	0·7	0·8	0·9	1·0
45	0·712	0·737	0·761	0·783	0·807	0·834	0·863	0·895	0·930	0·965	1·000
46	0·708	0·733	0·757	0·780	0·804	0·831	0·860	0·893	0·928	0·964	1·000
47	0·704	0·730	0·754	0·777	0·801	0·828	0·858	0·891	0·926	0·963	1·000
48	0·700	0·726	0·750	0·773	0·798	0·825	0·855	0·888	0·924	0·962	1·000
49	0·697	0·723	0·747	0·770	0·795	0·822	0·852	0·886	0·923	0·961	1·000
50	0·693	0·719	0·744	0·767	0·792	0·819	0·850	0·884	0·921	0·960	1·000
51	0·690	0·716	0·741	0·764	0·789	0·816	0·847	0·882	0·919	0·959	1·000
52	0·686	0·713	0·738	0·761	0·786	0·814	0·845	0·880	0·918	0·958	1·000
53	0·683	0·710	0·735	0·758	0·783	0·811	0·842	0·878	0·916	0·957	1·000
54	0·680	0·707	0·732	0·756	0·781	0·809	0·840	0·875	0·914	0·956	1·000
55	0·677	0·704	0·729	0·753	0·778	0·806	0·838	0·873	0·913	0·955	1·000
56	0·674	0·701	0·726	0·750	0·775	0·804	0·835	0·871	0·911	0·954	1·000
57	0·671	0·698	0·723	0·747	0·773	0·801	0·833	0·869	0·909	0·953	1·000
58	0·668	0·695	0·721	0·745	0·770	0·799	0·831	0·867	0·908	0·952	1·000
59	0·666	0·693	0·718	0·742	0·768	0·796	0·829	0·865	0·906	0·951	1·000
60	0·663	0·690	0·715	0·740	0·765	0·794	0·827	0·863	0·905	0·950	1·000
61	0·660	0·687	0·713	0·737	0·763	0·792	0·824	0·862	0·903	0·949	1·000
62	0·658	0·685	0·710	0·735	0·760	0·789	0·822	0·860	0·902	0·948	1·000
63	0·655	0·682	0·708	0·732	0·758	0·787	0·820	0·858	0·900	0·947	1·000
64	0·653	0·680	0·705	0·730	0·756	0·785	0·818	0·856	0·899	0·946	1·000
65	0·650	0·677	0·703	0·727	0·753	0·783	0·816	0·854	0·897	0·946	1·000
66	0·648	0·674	0·700	0·725	0·751	0·780	0·814	0·852	0·896	0·945	1·000
67	0·645	0·672	0·698	0·722	0·749	0·778	0·812	0·850	0·894	0·944	1·000
68	0·643	0·669	0·695	0·720	0·746	0·776	0·810	0·849	0·893	0·943	1·000
69	0·640	0·667	0·693	0·718	0·744	0·774	0·808	0·847	0·891	0·942	1·000
70	0·638	0·664	0·690	0·715	0·742	0·772	0·806	0·845	0·890	0·941	1·000
71	0·636	0·662	0·688	0·713	0·740	0·770	0·804	0·843	0·888	0·940	1·000
72	0·633	0·660	0·685	0·710	0·737	0·768	0·802	0·841	0·887	0·939	1·000
73	0·631	0·657	0·683	0·708	0·735	0·765	0·800	0·840	0·885	0·938	1·000
74	0·629	0·655	0·680	0·706	0·733	0·763	0·798	0·838	0·884	0·937	1·000
75	0·626	0·652	0·678	0·703	0·731	0·761	0·796	0·836	0·883	0·936	1·000
76	0·624	0·650	0·675	0·701	0·728	0·759	0·794	0·835	0·881	0·935	1·000
77	0·622	0·647	0·673	0·699	0·726	0·757	0·792	0·833	0·880	0·935	1·000
78	0·619	0·645	0·670	0·696	0·724	0·755	0·791	0·831	0·879	0·934	1·000
79	0·617	0·642	0·668	0·694	0·722	0·753	0·789	0·830	0·877	0·933	1·000
80	0·615	0·640	0·666	0·692	0·720	0·751	0·787	0·828	0·876	0·932	1·000
81	0·613	0·638	0·663	0·689	0·718	0·749	0·785	0·827	0·875	0·931	1·000
82	0·610	0·635	0·661	0·687	0·715	0·747	0·784	0·825	0·873	0·930	1·000
83	0·608	0·633	0·659	0·685	0·713	0·745	0·782	0·824	0·872	0·930	1·000
84	0·606	0·631	0·656	0·683	0·711	0·744	0·780	0·822	0·871	0·929	1·000
85	0·604	0·628	0·654	0·681	0·709	0·742	0·779	0·821	0·870	0·928	1·000
86	0·602	0·626	0·652	0·679	0·708	0·740	0·777	0·819	0·869	0·927	1·000
87	0·600	0·624	0·650	0·677	0·706	0·739	0·776	0·818	0·868	0·927	1·000
88	0·598	0·622	0·648	0·675	0·704	0·737	0·774	0·817	0·867	0·926	1·000
89	0·596	0·620	0·646	0·673	0·703	0·736	0·773	0·816	0·866	0·925	1·000
90	0·594	0·619	0·645	0·672	0·701	0·734	0·772	0·815	0·865	0·925	1·000

to experimental observations whilst Duncumb and Reed[10] gave the expression

$$J = Z[14(1 - \exp(-0 \cdot 1 Z)) + 75 \cdot 5 Z^{(-Z/7 \cdot 5)} - Z/(100 + Z)]$$

based on a large number of electron probe analyses of specimens of known composition. The latter is probably better for use with the simple ZAF correction procedure described here (with no allowance for continuum fluorescence) such as was used in making the original analyses to give the J values. These values of J are given in Table 7.2.

The calculation of this correction follows one of two routes: The simplifying assumptions often made are that Q is constant (and therefore cancels in the expression for Z) and since the ratio $S_A : S$ (standard) is a slowly varying function of E, S can be evaluated at an average value of E. In this case, for the purpose of substitution in a simplified equation for the atomic number correction Z, the equation

$$S = (\text{const}) \frac{Z}{A} \ln \left(\frac{1 \cdot 166 \bar{E}}{J} \right)$$

is evaluated for each element at a value of $\bar{E} = (E_0 + E_c)/2$ where E_c is the critical excitation potential for the radiation being measured. For the specimen, S is then calculated by a similar averaging process as was used for R, to give for the measurement of element A, $S_A = \sum_i S_{Ai} C_i$.

This method for obtaining an average value for S is perhaps logically the least satisfactory step in this simple method for calculating S; nevertheless, it is a very widely used procedure. Alternatively the integral can be evaluated numerically, and graphs to facilitate this have been given by Philibert and Tixier.[11] In practice, numerical integration using the logarithmic integral function is generally only incorporated into computer methods of making the correction, and then only rarely.

7.3.2 A, the Absorption Correction

The absorption correction,

$$A = \frac{f_A(\chi_A)}{f_{\text{Std}}(\chi(A))}$$

has to be evaluated for each element measured. $f_A(\chi_A)$ allows for absorption of radiation from element A in the specimen and $f_{\text{Std}}(\chi)$ the corresponding absorption in the standard. Different analytical expressions for $f(\chi)$ are available, based on different expressions for the ionization function $\phi(\rho z)$.

TABLE 7.2
Collected Data for Computing the ZAF Correction Factors

Element	Atomic number	Atomic weight	Critical Absorption Energies (keV)		Mean Ionization Potential J (keV)	Absorption Jump Ratio r		Fluorescence Yield	
			K series	L series		K	L	w_K	w_L
H	1	1·008	0·013 6						
He	2	4·003	0·024 6						
Li	3	6·940	0·055						
Be	4	9·02	0·116						
B	5	10·82	0·192			23·259		0·001	
C	6	12·01	0·283		0·246	21·739		0·002	
N	7	14·008	0·399		0·135	20·000		0·004	
O	8	16·00	0·431		0·127	18·518		0·007	
F	9	19·00	0·687		0·123	17·544		0·010	
Ne	10	20·183	0·874	0·022	0·123	16·393		0·014	
Na	11	22·997	1·08	0·034	0·126	15·625		0·020	
Mg	12	24·32	1·303	0·049	0·133	14·706		0·028	
Al	13	26·97	1·559	0·072	0·142	13·889		0·037	
Si	14	28·06	1·838	0·098	0·154	13·158		0·048	
P	15	30·98	2·142	0·128	0·166	12·500		0·062	
S	16	32·006	2·470	0·163	0·180	11·905		0·077	
Cl	17	35·457	2·819	0·202	0·194	11·364		0·095	
A	18	39·944	3·203	0·245	0·209	10·869		0·115	
K	19	39·096	3·607	0·294	0·224	10·417		0·138	
Ca	20	40·084	4·038	0·349	0·239	10·100		0·163	
Sc	21	45·10	4·496	0·406	0·255	9·709		0·190	
Ti	22	47·90	4·964	0·454	0·270	9·434			

TABLE 7.2—contd.

Element	Atomic number	Atomic weight	Critical Absorption Energies (keV)		Mean Ionization Potential J (keV)	Absorption Jump Ratio r		Fluorescence Yield	
			K series	L series		K	L	w_K	w_L
V	23	50·954	5·463	0·512	0·286	9·091		0·219	
Cr	24	52·01	5·988	0·574	0·301	8·849		0·249	
Mn	25	54·93	6·537	0·639	0·316	8·547		0·281	
Fe	26	55·85	7·111	0·708	0·332	8·333		0·314	
Co	27	58·94	7·709	0·779	0·347	8·065		0·347	
Ni	28	58·69	8·331	0·853	0·362	7·813		0·381	
Cu	29	63·54	8·980	0·933	0·377	7·576		0·414	
Zn	30	65·38	9·660	1·022	0·392	7·407		0·448	
Ga	31	69·72	10·368	1·117	0·407	7·194		0·480	
Ge	32	72·60	11·103	1·217	0·422	6·993		0·512	
As	33	74·91	11·863	1·323	0·437	6·849		0·542	
Se	34	78·96	12·652	1·434	0·451	6·667		0·572	
Br	35	79·916	13·475	1·552	0·466	6·494		0·600	
Rb	37	85·48	15·20	1·806	0·495	6·190	7·00	0·652	0·018
Sr	38	87·63	16·112	1·941	0·510	6·050	6·89	0·676	0·020
Y	39	88·90	17·042	2·079	0·524	5·930	6·78	0·698	0·023
Zr	40	91·22	17·998	2·22	0·538	5·813	6·67	0·719	0·025
Nb	41	92·91	18·987	2·374	0·553	5·714	6·54	0·739	0·028
Mo	42	95·95	20·002	2·523	0·567	5·58	6·45	0·757	0·033
Ru	44	101·70	22·127	2·838	0·595		6·24	0·789	0·036
Rh	45	102·91	23·234	3·000	0·609		6·14	0·804	0·039
Pd	46	106·7	24·347	3·172	0·623		6·06	0·818	0·043
Ag	47	107·88	25·517	3·352	0·637		5·952	0·830	0·047

Cd	48	112·41	26·712	3·538	0·651	5·88	0·842	0·051
In	49	114·76	27·928	3·729	0·665	5·78	0·852	0·054
Sn	50	118·7	29·190	3·928	0·679	5·714	0·862	0·059
Sb	51	121·76	30·486	4·132	0·692	5·617	0·871	0·063
Te	52	127·61	31·809	4·341	0·706	5·55	0·880	0·068
I	53	126·92	33·164	4·559	0·720	5·46	0·858	0·073
Cs	55	132·91	35·973	5·012	0·747	5·31	0·901	0·084
Ba	56	137·36	37·410	5·247	0·761	5·263	0·908	0·090
La	57	138·92	38·931	5·489	0·775	5·181	0·916	0·096
Ce	58	140·13	40·449	5·729	0·786	5·128	0·919	0·102
Hf	72	178·6	65·313	9·556	0·977	4·347	0·964	0·212
Ta	73	180·88	67·400	9·876	0·991	4·291	0·966	0·221
W	74	183·92	69·508	10·198	1·004	4·255	0·968	0·231
Re	75	186·31	71·662	10·531	1·017	4·201	0·970	0·240
Os	76	190·2	73·900	10·873	1·031	4·813	0·971	0·250
Ir	77	193·1	76·097	11·211	1·044	4·166	0·973	0·260
Pt	78	195·23	78·379	11·559	1·057	4·081 6	0·974	0·270
Au	79	197·2	80·173	11·919	1·071	4·032	0·975	0·280
Hg	80	200·61	83·106	12·285	1·084	4·000	0·976	0·291
Th	81	204·39	85·517	12·657	1·097	3·952	0·977	0·301
Pb	82	207·21	88·001	13·044	1·111	3·921	0·978	0·312
Bi	83	209·00	90·521	13·424	1·124	3·875	0·979	0·322
Th	90	232·04	109	16·3	1·217		0·985	0·396
U	92	238·03	115	17·2	1·244		0·986	0·417

Values of J after Duncumb and Reed,[10] of r after Heinrich's absorption data[14] and w after Burhop's equation (see text).

The most widely used is that due to Philibert[12]

$$f(\chi) = \frac{1 + h}{(1 + x/\sigma)(1 + h(1 + x/\sigma))}$$

where $\chi = \mu \csc \theta$ is a measure of the absorption of X-rays in the specimen or standard and is calculated for the absorption of the radiation from element A being measured, by summing for all the elements in the specimen, e.g. $\mu_{\text{alloy}}^A = \sum \mu_i^A C_i$ where C_i is the weight fraction of element in the specimen. σ, the Lenard coefficient, is a measure of electron absorption. Values were tabulated, in terms of incident electron energy E_0, by Philibert, but a more suitable value is best obtained from an equation proposed by Heinrich (see reference 4).

$$\sigma = \frac{4 \cdot 5 \times 10^5}{E_0^{1 \cdot 65} - E_c^{1 \cdot 65}}$$

where E_0 and E_c are in kiloelectron-volts. h is a parameter which depends on atomic number and allows for the variation of the shape of the $\phi(\rho z)$ curve with atomic number. The expression proposed by Philibert $h = 1 \cdot 2\,(A/Z^2)$, where A is atomic weight and Z is atomic number, was chosen to fit experimental $f(\chi)$ curves, and is the most widely used. There are different opinions as to how an average value of h should be calculated; $h = \sum h_i C_i$ is most widely used but

$$h = 1 \cdot 2 \frac{\sum A_i n_i}{(\sum Z_i n_i)^2}$$

also has some justification, where n_i is the atom fraction of element i. Although the different methods lead to different values of h, the difference in the overall calculation is very small and for most cases is insignificant.

In deriving the above formula, Philibert first produced the full expression

$$f(\chi) = \frac{1 + \{[h\chi\phi(0)]/[4\sigma(1 + h\phi(0)/4)]\}}{\left(1 + \dfrac{\chi}{\sigma}\right)\left(1 + \dfrac{h}{1 + h} \times \dfrac{\chi}{\sigma}\right)}$$

where $\phi(0)$ is the value of $\phi(\rho z)$ at the surface of the specimen. The second (shortened) version of the expression for $f(\chi)$ is obtained from this by taking $\phi(0) = 0$ (see Fig. 7.4), and this is the form that is in general use.

Bishop[13] proposed the expression

$$f(\chi) = \frac{1 - \exp(-2\chi\overline{\rho z})}{2\chi\overline{\rho z}}$$

where the mean mass depth of X-ray generation is deduced from the Philibert model as

$$\overline{\rho z} = \frac{1 + 2h}{\sigma(1 + h)}$$

Bishop's function is based upon the 'rectangular' approximation where $\phi(\rho z) = $ constant for the range $0 < \rho z < 2\overline{\rho z}$ as shown in Fig. 7.4. This gives surprisingly good results and is probably better than the usual Philibert equation for cases in which long wavelengths and consequently high absorption are concerned. If the absorption is very high it is only X-rays produced from the near-surface layers of the specimen that emerge and use of the approximation $\phi(0) = 0$ is clearly undesirable in this case.

Values of mass absorption coefficient μ are available in a very convenient form compiled by Heinrich[14] for elements with atomic number higher than sodium (see Table 7.3). For the longer wavelengths from elements with $Z < 11$ the experimental data are poor and limited. Perhaps the most reliable direct measurements are due to Henke et al.[15] whilst probably the best available data derived from microanalysis measurements of boron, carbon, nitrogen and oxygen are given in the work of Ruste and Gantois[16] and of oxygen by Love et al.[17] If the absorption coefficient is very large, particularly in the neighbourhood of an absorption edge, then uncertainty in the analysis is due to both errors in the value used for the absorption coefficient and the approximations made in calculating the correction. For this reason it is often recommended that a value of $f(\chi) < 0.8$ should not be used, but that the voltage E_0 should be reduced to make the magnitude of the absorption correction less significant—if values of $f(\chi) < 0.7$ are used significant absorption errors must be expected.

For non-normal incidence, as is used in many scanning electron microscopes fitted with EDX analytical facilities, the distribution of X-ray production with depth $\phi(\rho z)$ depends upon specimen tilt. As the angle β between electron beam and specimen surface is reduced, the mean depth of X-ray production moves closer to the specimen surface and the absorption correction is reduced.

The simplest way of modifying the correction is to multiply $\overline{\rho z}$ (or χ) by $\sin \beta$, on the assumption that the distribution of X-ray production along the axis of the beam is independent of β. From Monte Carlo calculations it has been suggested that this is not so, and that a better factor to use would be $(1 - 0.5 \cos^2 \beta)$. It is probably best to use this latter factor, and ensure that $\beta \geq 50°$ as a compromise to minimise the uncertainty in cases where the correction for non-normal incidence has to be made.

TABLE 7.3

Mass absorption coefficients, cm^2/g (after Heinrich)[14]

Absorber	Emitter, wavelength in A									
	Al 8·337	Si 7·126	Ti 2·748	V 2·504	Cr 2·291	Mn 2·103	Fe 1·937	Co 1·790	Ni 1·659	Cu 1·542
13 Al	385·7	3 493·2	247·0	190·8	149·0	117·4	93·4	75·0	60·7	49·6
14 Si	503·4	327·9	304·3	235·2	183·8	145·0	115·5	92·8	75·2	61·4
22 Ti	2 288·0	1 490·6	110·6	85·8	597·0	472·5	377·5	304·4	247·3	202·6
23 V	2 621·4	1 707·8	126·7	98·3	77·1	531·1	424·3	342·1	278·0	227·7
24 Cr	3 000·5	1 954·8	145·0	112·5	88·2	69·9	474·2	382·3	310·7	254·4
25 Mn	3 415·6	2 225·3	165·1	128·1	100·5	79·5	63·5	422·5	343·6	281·6
26 Fe	3 840·6	2 502·1	185·6	144·0	113·0	89·4	71·4	57·6	379·6	311·1
27 Co	4 330·8	2 821·6	209·3	162·4	127·4	100·8	80·6	64·9	52·8	341·2
28 Ni	4 837·5	3 151·6	233·8	181·4	142·3	112·6	90·0	72·5	58·9	48·3
29 Cu	5 376·8	3 503·0	259·8	201·6	158·1	125·2	100·0	80·6	65·5	53·7

The table gives data for the absorption of $K\alpha$ radiation from some selected elements. The complete table covers the range Na–Mo (K radiation), Ga–U (L radiation) and Yb–U (M radiation).

7.3.3 F, the Fluorescence Correction

I_{jA}^f is the fluorescent radiation from element A emitted from the specimen, produced as a consequence of absorption of characteristic radiation from element j. Such a contribution will only arise if the energy of the radiation from j is greater than the energy (E_c) of the absorption edge of element A that is appropriate to the radiation being measured. This can be ascertained from tables of emission and excitation energies (or wavelengths).

To calculate γ_{jA}, the ratio of I_{jA}^f to the intensity that is directly excited, a formula originally derived by Castaing and slightly modified by Green and Cosslett[18] is used. The formula is somewhat involved, and it is necessary to appreciate the significance of each term so that correct values can be assigned to the different parameters. The derivation is straightforward and the principal steps are outlined below:

If for simplicity it is assumed that the primary radiation is generated on the surface, half will escape from the specimen and half will be absorbed in the specimen, $0 \cdot 5 I_j$.

A fraction $C_A \mu_A^j / \mu^j$ will be absorbed by A atoms in the specimen where μ_A^j is the mass absorption coefficient for the radiation j in pure A and μ^j is the mass absorption coefficient for the radiation j in the specimen. A fraction of this absorbed radiation will give rise to ionization of the appropriate shell in the A atoms, and a value for this is obtained by considering the absorption on either side of the edge. If r_A is the ratio of absorption coefficients on either side of the relevant absorption edge for element A, then a fraction $(r_A - 1)/r_A$ of the absorbed radiation will give ionizations in the appropriate shell. Values of r derived from Heinrich's absorption data are given in Table 7.2. The probability of an ionized atom giving an X-ray emission is the fluorescence efficiency ω_{AK} for the K shell of element A. Combining these factors gives the intensity of characteristic radiation produced by fluorescence

$$I_{jA}^f = 0 \cdot 5 \, C_A \frac{\mu_A^j}{\mu^j} \frac{r_A - 1}{r_A} \omega_{AK} I_j$$

where I_j is the intensity of radiation from element j that is causing fluorescence. The intensity I_j depends upon the product of several terms: the weight fraction of j atoms present in the specimen C_j, the fluorescence efficiency ω_{jK}, the reciprocal atomic weight $1/A_j$, a term in E_0 and E_c given by Green and Cosslett[18] as $(U_j - 1)^{1.67}$, and a constant P_K or P_L which allows for the different efficiency of production for K and L series spectra.

Combining these terms gives

$$I_j = C_j P_K \omega_{jK} (U_j - 1)^{1.67} / A_j$$

and using a similar expression for I_A gives the required form of the correction factor

$$\frac{I_{jA}^f}{I_A} = 0 \cdot 5 \, C_j \, P_{jK} \frac{\mu_A^j}{\mu^j} \frac{r_A - 1}{r_A} \omega_{jK} \frac{A_A}{A_j} \left(\frac{U_j - 1}{U_A - 1} \right)^{1.67}$$

where the factor P_{jK} which refers to j radiation producing fluorescence of the measured K radiation from element A, allows for the relative intensities of K and L radiation if the radiations involved come from different series. Thus for $K \to K$ and $L \to L$ fluorescence $P_{KK} = P_{LL} = 1$, whilst for K exciting L, $K \to L$, the accepted value is $P_{KL} = 0.25$ and for $L \to K$, $P_{LK} = 4.0$.

Values for ω_j are given in Table 7.2, and are based on Burhop's equation: $\omega = Z^4/(a + Z^4)$ where $a_K = 10^6$ and $a_L = 10^8$. More recent data are given by Burhop and Asaad[19] but the changes are not very significant. Normally little error is introduced by neglecting the spread of wavelengths in the series of lines (K or L spectra) causing fluorescence, and it is normally considered that all the intensity is produced at the wavelength of the most intense line ($K\alpha$ or $L\alpha$) causing fluorescence.

Otherwise we must multiply the fluorescence efficiency ω_j by the quantum weighting ξ_j of each line causing fluorescence and sum over all such lines in the spectrum of element j, using the appropriate values for μ^j.

To allow for absorption of I_{jA}^f it is first necessary to know the distribution with depth, and this is obtained by calculating the attenuation of the radiation I_j with depth. Assuming that directly excited radiation is generated at a point on the surface, then if dI^f is produced at depth z in a layer of thickness dz, this will be attenuated as it emerges from the specimen by the factor $\exp(-\mu_A \rho z \operatorname{cosec} \theta)$. A double integration over the volume of the specimen then gives a factor $\ln(1 + x)/x$ where $x = \mu^A \operatorname{cosec} \theta/\mu^j$ to allow for absorption in the alloy. If the primary radiation is considered to be produced over a finite depth in the alloy and is assumed to follow an exponential distribution $I_j(z) = I_j(0) \exp(-\sigma \rho z)$ where σ is the Lenard coefficient, then this gives an additional term in the correction $\ln(1 + y)/y$ where $y = \sigma^j/\mu^j$ and $\sigma = 4.5 \times 10^5/(E_0^{1.65} - E_c^{1.65})$

The complete correction formula for fluorescence is then

$$\frac{I_{jA}^f}{I_A} = 0.5 \, C_j \frac{\mu_A^j}{\mu^j} \frac{r_A - 1}{r_A} \times \omega_{jK} \frac{A_A}{A_j} \left(\frac{U_j - 1}{U_A - 1} \right)^{1.67} \times \left\{ \frac{\ln(1 + x)}{x} + \frac{\ln(1 + y)}{y} \right\}$$

A parameter $\gamma_{jA} = I_{jA}^f/I_A$ has to be evaluated for each radiation that can cause fluorescence, and the sum of these terms over all j included in the correction formula.

7.4　HAND CALCULATION OF THE ZAF CORRECTION

If hand calculations are to be used, iteration is very tedious and it is often better to compute K values for chosen concentrations and then use graphical methods. In addition it is often adequate to make an estimate (using experience) of the magnitude of the various terms Z, A and F, and only compute those which are significant. Two examples of such calculations are given below:

To calculate K_{Cu} and K_{Al} for a Cu 10 wt% Al alloy for $E_0 = 20 \, kV$ and a take-off angle $\theta = 75°$.

We note that the correction will be in Z and A, there is no F correction (ignoring any white fluorescence effect).

The Z correction—a table of parameters which are independent of specimen composition is compiled:

	Cu	Al
Z	29	13
A	63.5	27
E_c, keV for element measured	8.98	1.56
$\bar{E}$, keV for element measured	14·49	10·78
σ, using appropriate E_c for element measured	4378	3259
$1/U$, using appropriate E_c for element measured	0·45	0.08
J, keV	0·377	0.142

Secondly an array of S values is compiled for each radiation measured for each element present, and the appropriate mean value for the specimen computed. This is repeated for values of R.

S array

	Cu	Al	Cu10Al
Al Kα	1·602	2·159	1·658
Cu Kα	1·737	2·301	1·793

R array

	Cu	Al	Cu10Al
Al Kα	0·810	0·922	0·821
Cu Kα	0·883	0·957	0·890

Hence

$$K_{Cu} = 0·9 \times \frac{0·890}{0·883} \times \frac{1·737}{1·793} = 0·879 \qquad (\text{i.e. } \sim 2\% \text{ correction})$$

and

$$K_{Al} = 0·1 \times \frac{0·821}{0·922} \times \frac{2·159}{1·658} = 0·116 \qquad (\text{i.e. } \sim 16\% \text{ correction})$$

The A correction—a table of mass absorption coefficients is compiled for each radiation used for each absorber present, i.e. for each standard and for the specimen, also values of h. An $f(\chi)$ value is then computed for each standard and values for the specimen are also calculated (see Table 7.3).

	Cu	Al	$Cu10Al$
$\mu(Al\,K\alpha)$	5376·8	385·7	4877·7
$\mu(Cu\,K\alpha)$	53·7	49·6	53·3
$h = 1\cdot2A/Z^2$	0·092	0·192	0·102
$\chi(Al\,K\alpha)$	5566·5	399·3	5049·8
$\chi(Cu\,K\alpha)$	55·6	51·3	55·1
$f(\chi)(Al\,K\alpha)$	—	0·8736	0·3430
$f(\chi)(Cu\,K\alpha)$	0·9864	—	0·9864

Hence

$$K_{Cu} = 0\cdot879 \times \frac{0\cdot9864}{0\cdot9864} = 0\cdot879 \qquad \text{(no absorption correction)}$$

and

$$K_{Al} = 0\cdot116 \times \frac{0\cdot3430}{0\cdot8736} = 0\cdot0455$$

(note that the absorption correction is $\sim 250\%$ and good accuracy should not be expected)

To calculate K_{Co} *for a Cu10 wt% Co alloy for* $E_o = 20\,kV$ *and a take-off angle of* $\theta = 75°$.

Note that the correction will be almost entirely in F (fluorescence due to excitation of the Co K-shell by Cu K radiation), and that the Z correction will be small, since $Z(Co) \approx Z(Cu)$, and the A correction will also be small since mass absorption coefficients μ for Co Kα in Co $\approx \mu$ for Co Kα in Cu (Table 7.3) and the values are not large.

We will calculate the F correction alone, and assume that all the radiation from the K shell of Cu acts as though it had the wavelength of Cu Kα.

A table of parameters that are independent of composition is compiled using data from Table 7.2.

$$
\begin{aligned}
&\omega\,(Cu) = 0\cdot414 & &E_c\,(Co) = 7\cdot709 \\
&E_c\,(Cu) = 8\cdot980 & &U\,(Co) = 2\cdot594 \\
&U\,(Cu) = 2\cdot277 & &\sigma\,(Co) = 4050\cdot1 \\
&\sigma\,(Cu) = 4378\cdot2 & &A\;Co = 58\cdot9 \\
&A\,(Cu) = 63\cdot5 & &r\,(Co) = 8\cdot065
\end{aligned}
$$

P_{KK} for Cu $K \to$ Co $K = 1$

An array of mass absorption data is then compiled from Table 7.3

	Cu	Co	$Cu10Co$
$\mu(\text{Cu K}\alpha)$	53·7	341·2	82·4
$\mu(\text{Co K}\alpha)$	80·6	64·9	79·0
$\chi(\text{Cu K}\alpha)$	55·6	353·2	85·3
$\chi(\text{Co K}\alpha)$	83·4	67·2	81·8

$$x = \frac{81·8}{82·4} = 0·9927 \qquad y = \frac{4378·2}{79·0} = 55·420$$

$$\therefore \quad \frac{\ln(1 + x)}{x} + \frac{\ln(1 + y)}{y} = 0·7673$$

$$\therefore \quad \gamma(\text{Cu}) = 0·5 \times 0·414 \times 0·876 \times \frac{58·9}{63·5} \times \frac{1·227}{1·594}$$

$$\times 1·65 \times \frac{341·2}{82·4} \times 0·7673 \times 0·9$$

$$= 0·3470 \times 0·9 = 0·312$$

$$\therefore \quad K_{\text{Co.}} = 0·1 \times 1·312 = 0·131 \qquad \text{(a 30\% correction)}$$

7.5 COMPUTER CALCULATION OF THE ZAF CORRECTION

At least a hundred different programs have been written to compute the ZAF correction. In addition to easing the computational effort many of these programs retrieve much of the necessary data either from large data files or by the use of approximate formulae. They also can give warnings about an unsuitable value of E_0 or other operational variable.

Some of the more sophisticated versions e.g. COR2[20] or EMPADR VII[21] require the use of a large computer. They may perform numerical integration when calculating values of S and also when making allowance for fluorescence caused by continuum radiation. An example of a simpler version is FRAME[22] which requires only 4000–5000 words of memory and can be used on a desk-top computer. With this it is an economic possibility to use a dedicated mini-computer on-line with the microanalyser to give immediate analytical results, in which case corrections for dead-time and background can conveniently be included. The difference in accuracy between sophisticated and simple programs does not seem to be significant, although the simpler versions do require rather more stringent working conditions for satisfactory application.

When a computer is used the calculation usually proceeds from a measured value of K to a value of concentration C. Several components may be measured at once, and for each one the corresponding Z, A and F terms require values of C for each component before they can be calculated. Simple iteration using the formula $K_A^m = (ZAF) C_A$ does not necessarily lead to rapidly convergent values for C_A, C_B etc., especially if large absorption or fluorescence corrections are involved. A hyperbolic approximation to the K versus C relation for each element is justified by experiment and is often used in the iterative process, viz:

$$\frac{1 - K_A}{K_A} = \alpha \frac{1 - C_A}{C_A}$$

where α is a constant relating to the analysis of element A. The first value for C_A is obtained using the approximation $C_A' = K_A^m / \sum K_i^m$, that is the intensity ratios for this first approximation are normalized to render the sum of approximate mass fractions equal to unity. The true intensity ratio for this concentration is then calculated $K_A' = (ZAF)C_A'$, used in the hyperbolic relation to give a value of α.

This value of α, together with the original un-normalized value of K_A^m, are then used in the hyperbolic relation to give a better approximation C_A''; that is

$$C_A'' = \frac{\alpha K_A^m}{1 - (1 - \alpha) K_A^m}$$

Further iteration uses C_A'' and K_A^m to give a better value of α with which to get a better value for C_A. At this stage of the process it is better not to normalize either K_A^m or the approximations to C_A in case there are experimental errors, or the presence of an element has been overlooked, in which case the final values of mass fraction will fail to sum to unity.

7.6   AN EMPIRICAL CALIBRATION METHOD

A method developed by Ziebold and Ogilvie[2] requires the use of specimens of known composition to develop a calibration curve. The form of this is based on the hyperbolic relationship mentioned below:

$$\frac{1 - K_A}{K_A} = \alpha_A \frac{1 - C_A}{C_A}$$

which can be written as $C_A/K_A = \alpha_A + (1 - \alpha_A)C_A$. Since a linear relation between C_A/K_A and C_A is found experimentally for a binary system, the constancy of the factor α_A is justified by experiment. This equation is then used as a convenient way of incorporating data obtained from several calibration specimens into an analytical system, and can readily be extended to a ternary or more complicated alloy if it is assumed that α_A is a weighted mean of α-values corresponding to each of the n elements present in the specimen. Thus for measuring element A,

$$\alpha_A = \sum_i \alpha_{Ai} \frac{C_i}{1 - C_A}$$

The success of the technique depends on the constancy of the factors α_A, α_B, etc. for which there is some experimental rather than fundamental justification, but for a small range of composition, and given suitable calibration specimens of similar composition, there is no doubt that the method is capable of better accuracy than direct reference to pure standards using the ZAF correction. However, the preparation of homogeneous alloys is often more difficult than expected, and of course the method depends directly on the accuracy of the chemical analysis.

7.7   ANALYSIS USING AN ENERGY DISPERSIVE DETECTOR

The components of an energy dispersive spectrometer (EDX) and the physical basis of its operation have been discussed in Chapter 6. Such equipment can be used to give quantitative analysis in a very similar way to that described previously for a crystal wavelength dispersive spectrometer (WDX), in that the intensities of the characteristic lines emitted by specimen and standard can be compared, and from this ratio the concentration of each element present can be computed. The ZAF correction is unaffected by the method of X-ray detection that is used.

However, with a Si(Li) detector the relatively poor resolution compared with a crystal spectrometer means that problems due to overlapping lines are more serious, and likewise a satisfactory method of subtracting the background level from each characteristic line is equally important. In addition a crystal spectrometer measures the intensity of only one wavelength (depending upon the Bragg angle) at any one instant, whilst in an EDX the whole spectrum is measured simultaneously in the form of a pulse height spectrum. Consequently high count rates are involved and this, coupled with a long dead-time necessary to get good energy resolution, means that dead-time losses and other effects associated with getting a true pulse height spectrum corresponding to the energy spectrum of the emitted X-rays, pose a significant problem. (These experimental aspects of the use of the EDX for quantitative work are given in Chapter 6 and have been discussed in more detail by Woldseth[23] and Beaman and Isasi.[4])

The use of a small computer in conjunction with the EDX is essential. This can be programmed so that much of the manipulation and decision-making associated with making an analysis is carried out automatically, both for processing the spectrum to obtain intensities in the various X-ray lines, and computing the relation between intensity and concentration. In the first instance all the lines in the spectrum must be identified, and this is done either manually (and subjectively for very weak lines) using a visual display of the spectrum with a computer-aided line identification display, or it can be done automatically, searching for spectral lines of known shape and width corresponding to the resolution of the detector. Probably some combination of these two techniques is desirable. Knowledge of the relative intensities (that is the quantum weighting ξ) of the lines present in the spectrum of a given element can be used in order to make allowance for overlapping lines in the spectrum obtained. One method for allowing for overlapping lines, and at the same time allowing for the background intensity, is to subtract the spectrum corresponding to each of the elements present, until a background corresponding to the white spectrum from the specimen is obtained. Whatever system is used, some degree of flexibility in operation is desirable, since in some cases rapid processing of a complicated spectrum will be wanted, whilst on other occasions the best possible precision in processing a relatively simple spectrum will be required.

Simultaneous detection of the complete spectrum gives the opportunity of using an EDX in a way which does not require reference to standards. The procedure adopted is such that if specimens of similar known composition are available they can readily be incorporated so as to calibrate the method and so give a more accurate analysis. However, without the availability of such standards, the method involves measuring the characteristic intensities of lines corresponding to each element present in the specimen. Matrix effects must then be allowed for, using

$$I_{\text{emerging}} = I_{\text{produced}} \,(ZAF)$$

where the factors Z, A and F are calculated for the specimen as described earlier. Allowance must also be made for the efficiency (as a function of energy) of X-ray *production* and for the efficiency of X-ray *detection*. The efficiency of a WDX system is somewhat variable and frequent reference to standards is necessary. The efficiency of an EDX system is very stable and so a once-and-for-all (or at least infrequent) calibration to give the efficiency of production and detection of different wavelengths, is acceptable.

X-ray production depends upon the efficiency of production of ionized atoms (a function of E_0 and E_c) and the efficiency with which these produce radiation. Following the expression used when calculating the characteristic F correction factor, the intensity produced directly from an element of concentration C_i, atomic weight A_i is:

$$I_{iK} = C_i P_K \omega_{iK} \xi (U_i - 1)^{1.67} / A_i$$

where P_K is a factor to allow for the efficiency of excitation in the K shell, whilst a similar factor P_L refers to L radiation.

The efficiency of detection depends upon absorption in the detector so as to produce electron-hole pairs, and likewise absorption in the detector window, the inactive surface layer of the detector and any other material which may prevent the quantum reaching the detector. Due to this absorption the intensity I_{iK} will be reduced by $\exp(-\sum \mu_j \rho_j x_j)$ where $\rho_j x_j$ is the mass thickness of one of the absorbing layers and μ_j is the mass absorption coefficient for this layer. For a given radiation from an element, all these factors, except for the term in over-voltage ratio U, are constant, and the efficiency of the system for K radiation and for L radiation is a parameter which is expected to be a smooth function of atomic number.

Alternatively we can write

$$C_A = \alpha_A (Z_A A_A F_A)^{-1} I_A^m / \sum \alpha_i I_i^m$$

$$C_B = \alpha_B (Z_B A_B F_B)^{-1} I_B^m / \sum \alpha_i I_i^m, \text{ etc.}$$

and the factors α_A, α_B, etc., are obtained from a calibration specimen, or by reference to pure element standards.

A comparative study of the techniques used for quantitative analysis using a Si(Li) detector has been given by Statham.[24] For the analysis of thin foils, which are used to obtain better spatial resolution than is possible with solid specimens, a similar procedure is used, since variation in foil thicknesses is an uncontrolled variable. Provided the foil is thin enough, absorption can be neglected, and often any ZAF matrix correction is omitted and all correction is embraced in the α parameters.

7.8 LIGHT ELEMENT ANALYSIS

Light elements are usually defined as those with atomic numbers less than 9 (fluorine). The characteristic radiation produced has a relatively long wavelength and this is strongly absorbed in the specimen leading to a large absorption correction; there is good reason to use standards which are of similar chemical composition to the specimen. The shape and position of these long-wavelength spectral lines is sensitive to chemical combination and consequently there is also an advantage in using standards where the measured element is in a similar chemical state to that of the specimen.

The ZAF correction is no different in principle from that in normal use, but the usual form of the A term (which is the predominant correction) is likely to be inadequate. This is because only a small proportion of the total characteristic radiation generated, that from the near-surface layer, can emerge; consequently an $f(\chi)$ function based on a good approximation for near-surface ionization is required. From Fig. 7.4 it can be seen that the function used for the simple Philibert correction deviates strongly from the true ionization function near the surface, and the simple rectangular approximation may well be more appropriate for light element analysis. In order to calculate the mean mass depth of generation $\bar{\rho z}$, values of h (to give atomic number dependence) and σ (to give variation with E_0) are required. For the reason described above the expressions used with the simple Philibert correction are inappropriate and values fitted to a large number of

TABLE 7.4

Mass absorption coefficients (cm^2/g) for O, N and C K radiation, in some selected absorbers (after Ruste and Gantois[16])

Absorber	Z	Emitter K radiation wavelength in Å		
		O 23·6	N 31·6	C 44·7
C	6	11 197	25 441	2 270
N	7	17 519	1 634	3 836
O	8	1 250	2 585	6 041
Al	13	6 515	13 705	29 100
Si	14	8 383	17 678	35 800
Cr	24	3 343	5 762	10 906
Mn	25	3 801	6 591	12 612
Fe	26	4 300	7 500	14 502
Co	27	4 841	8 493	16 587
Ni	28	5 427	9 574	18 880
Cu	29	6 060	10 747	21 392

measurements by Love *et al.*[25] are probably the best to use:

$$\sigma = 7.05 \times 10^5/(E_0^{1.77} - E_c^{1.77}) \quad \text{and} \quad h = 1.06A/Z^2$$

with the suggestion that for a multicomponent specimen the average value $h = 1/\sum(C_i/h_i)$ should be used. Alternative expressions for σ and h have been proposed by Ruste and Gantois,[16] but these are based on more limited data.

One of the biggest uncertainties in the analysis of light elements is due to the lack of reliable data for mass absorption coefficients at long X-ray wavelengths. Sources of data are Gray and Wert,[26] Henke and Elgin[27] and Ruste and Gantois.[16] The latter are derived from microanalysis measurements, and some useful values (for which the different sources agree to within a few per cent) are given in Table 7.4.

7.9 EXPERIMENTAL ERRORS AND THEIR CONTROL

In measuring $I_A/I(A)$ there are four measurements to make—the peak and background intensity on specimen and standard. If the aim is to make an analysis with an accuracy of, say, 1 % of content, then the measurement of I_A and $I(A)$, and the constancy of conditions during the measurements must be rather better than this, say within about 0.5 % of content. Hence the error in any one intensity measurement due to statistics or other random error, or due to any systematic error, must be less than this.

7.9.1 Statistical Errors

Statistical errors due to the random nature of quantum production are easily dealt with since if a total of N counts are measured (in a time t to give a count rate of N/t cps) then the standard deviation of the measurement of N is $\sqrt{N}$ for large values of N. This can be applied to all measurements, and is an inevitable error associated with the random nature of the production of quanta. There are other sources of random error which lead to a scatter of measurements about a mean value; n measurements of N_i for $i = 1$ to n enable us to obtain the mean value $\bar{N} = \sum N_i/n$ and an estimate for the standard deviation $S = [\sum(N_i - \bar{N})^2/(n-1)]^{1/2}$.

A value of $N = 100\,000$ is required in order to achieve a $\sqrt{N}$ value corresponding to an error of 0.32 %; $N = 10\,000$ corresponds to a standard deviation of 1 %.

The counting strategy that is generally adopted is to use a preset counting time of between 10 and 500 s. Times greater than 500 s are inconvenient, and increase the demand on the overall stability of the instrument, but may be used for very low counting rates. The total time should be divided between peak and background measurements in such a way as to minimise the overall error due to counting statistics. Equal times for peak and background are appropriate if the intensities are virtually equal, as when trace elements are being determined; otherwise the ratio of the counting times should be the square root of the peak-to-background ratio.

A useful check on the overall performance of the instrument is to make $n \approx 10$ measurements of the ratio K for a known *homogeneous* specimen, making due allowance for the necessary background and dead-time corrections. The mean value obtained is then $\bar{K} = \sum K_i/n$ and the fractional standard deviation is

$$[\sum (K_i - \bar{K})^2/(n - 1)]^{1/2}/\bar{K}$$

For an instrument operating satisfactorily this should not be greater than three times the corresponding error associated with the statistical nature of the counting process

$$[(P + B)/(P - B)^2 + (P_s + B_s)/(P_s - B_s)^2]^{1/2}$$

where P and B are the total counts accumulated for peak and background on the specimen, and the subscript s denotes the same quantities for the standard.

Systematic errors due to some consistent procedural or instrumental error are more difficult to locate and eliminate. There are some common sources of such errors in microanalysis, and important cases will be considered below:

7.9.2 Non-linearity of X-ray Counting Equipment
The non-linearity of X-ray counting equipment normally takes the form of an increasing loss of counts as the count rate (cps) is increased. Figure 7.5 shows the expected relationship between true count-rate (which can be taken to be directly proportional to probe-current) and the observed count rate.

There are several possible reasons for this non-linearity:

Dead-time losses are always present. They are due to the counting equipment being effectively 'dead' for a short time, the dead-time τ, after having received a count. If another quantum arrives during this period it will not be detected, and consequently as the count rate is increased, a bigger fraction of the quanta will arrive within time τ from the previously

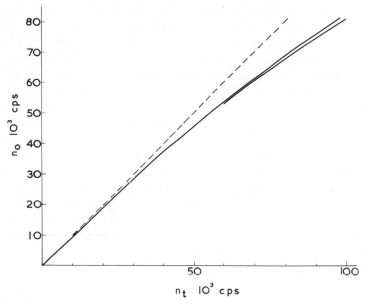

Fig. 7.5. Relation between true count rate n_t and observed count rate n_0 for a dead-time of $3\,\mu s$. Upper line—due to non-extendible dead-time (approaches a value of $3\cdot3 \times 10^5$ cps). Lower line—due to extendible dead-time.

detected quantum and will consequently not be counted. Dead-times are associated with different sections of the counting equipment. Non-extendible dead-times relate to equipment which completely ignores the existence of any non-counted pulses. The equipment observes n_0 cps and is 'dead' for a time $n_0\tau$ each second. The 'live-time' is $1 - n_0\tau$ and the true counting rate is given by

$$n_t = n_0/(1 - n_0\tau)$$

Extendible dead-times relate to equipment where the initiation of a dead period occurs at the arrival of each quantum, whether it is detected or not. Since the interval between random pulses follows a Poisson distribution, and the fraction of intervals greater than τ is given by $\exp(-n_t\tau)$

$$n_0/n_t = \exp(-n_t\tau)$$

These two different kinds of dead-time are applicable to different parts of the equipment and so the overall correction is often complicated. Fortunately, provided the correction is less than 10%, the difference between the two types is insignificant, and since the non-extendible correction involves the simpler arithmetic, it is this that is usually used with a maximum 10% correction limit on its application.

Pulse height depression is frequently observed and can arise in the counter or the amplifier. It is a phenomenon where the pulse size decreases (approximately linearly) with count rate, and consequently if the pulses are initially only just bigger than the discriminator level (which they must exceed in order to be counted) then an increasing fraction will be lost as the count rate is increased. This leads to a similar loss as that due to dead-time, but it varies with count rate in a different way, and it can usually be eliminated, at least in part, by increasing the size of the pulses. Errors due to this effect occur most frequently when pulse height analysis is being used.

Measurement of the dead-time τ can conveniently be carried out by plotting count rate for a particular X-ray line as a function of probe-current. This method depends upon a direct proportionality between true count rate n_t and probe current, $n_t = ki$, when substitution in the equation for non-extendible dead-time gives

$$\frac{n_0}{i} = k - k\tau n_0$$

For a constant dead-time a straight line will be found by plotting n_0/i versus n_0 from which a value for τ can be obtained (Fig. 7.6). The method depends upon having a reliable and precise measurement of the probe current.

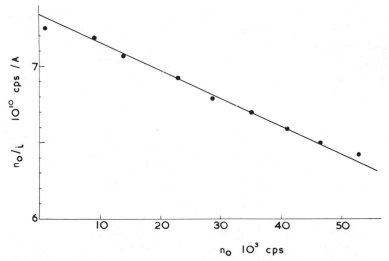

Fig. 7.6.   Graph used to compute a value for the dead-time giving intercept k = counts per sec. per μA and from the slope a dead-time $\tau = 2 \cdot 5\,\mu$s.

7.9.3 Re-setting Errors

Re-setting errors occur whenever a control is readjusted. The magnitude of the error can always be found by direct experiment leaving other control settings fixed. In WDX microanalysis this error often applies to spectrometer re-setting, and always applies to focussing in the optical microscope so as to set both specimen and standard at the same height and on the focus of the X-ray crystal spectrometers. These two adjustments are interdependent in that if the spectrometer is set just off the peak of the spectral line, then the measured intensity will be very sensitive to specimen height and re-setting errors due to optical focussing will be serious, in addition to those resulting from spectrometer re-setting. Careful setting of the spectrometer on the line peak when the source of X-rays has previously been located accurately on the focussing circle of the spectrometer will minimise such errors.

7.9.4 Specimen-associated Errors

Errors associated with the specimen are particularly prevalent for non-flat, thermally unstable materials. Variation in take-off angle θ is serious if there is strong absorption, and small tilts in the surface of the specimen, especially analysis from the sides of depressions or holes, can give results having large errors. Instruments where the take-off angle is small are particularly critical and difficult to use because of the increased sensitivity to this feature. Thermal instability of the specimen is found in many crystalline compounds, glasses and geological specimens. A useful check is to plot the intensity of each wavelength being measured as a function of specimen current. A linear relation is expected (except for curvature at high count rates because of dead-time effects) and the onset of instabilities related to the specimen can generally be found at a given specimen current, which is then the limiting specimen current for that material. Defocussing the electron probe is generally found to raise this limiting value.

Surface finish of the specimen is generally satisfied by a normal metallographic polish. Cases where absorption in the specimen is high are the most critical and the magnitude of the error is easily assessed by making small displacements of the electron probe in a region where the composition is known to be uniform. If the trouble is due to fine scratches then rotation of the specimen to set the direction of the scratches directly towards the spectrometer will minimize the effect. Etching the specimen often leads to the formation of surface films which, particularly in specimens containing two or more phases, can lead to misleading results. Sometimes a surface film contains a contaminant from the etching solution, and any form of etching or chemical treatment of the surface should be avoided. Similar trouble can arise because of contamination built up on the specimen where the electron probe impinges, and again the importance of such contamination can be assessed by moving the probe to a neighbouring uncontaminated area of the specimen. The contamination generally contains a high concentration of carbon and comes from oil vapour present in the vacuum. It is of course particularly serious if analysis for carbon is required, and such contamination can be reduced to a low level by the use of an effective cold finger in the neighbourhood of the specimen in cases where this proves necessary.

Non-conducting specimens require coating with a layer of carbon, aluminium or copper in order to provide a path for the electrons and prevent charging of the specimen. Vacuum evaporation or sputtering is usually used to deposit the layer. In many cases only about 200 Å of conducting layer is required and this may have no significant effect on the measured intensity. However, for longer wavelength (highly absorbed) radiation, or the use of low accelerating voltages, it is necessary to control the thickness of the layer that is deposited very carefully. Simultaneous deposition on specimen and standards is frequently used to ensure constant layer thickness.

7.9.5 Location of the Analysis

Location of the analysis on some microstructural feature is obtained in different ways depending on the construction of the instrument. If simultaneous electron bombardment and visual observation are available, then use of a specimen showing cathode luminescence locates the electron probe relative to a cross-wire or graticule in the microscope eyepiece. In this way the analysis can be located to within a few micrometres on specimens which do not exhibit luminescence. To improve on this, reference to a scanning electron image (at variable and higher magnification) is generally employed. Instability of the position of the probe on the specimen, when it is supposed to be stationary, can lead to errors in analysis in specimens with fine structure. This can be caused by warping of the filament or more often by charging effects, where dirt in the column or non-conducting parts of the specimen are charged and repeatedly discharged. This can be directly observed if the specimen shows luminescence under the electron beam, or it can be inferred from fluctuations in the specimen current and/or emitted X-ray intensity. The position and shape of the contamination mark that forms where the electron probe impinges on the specimen is also useful in relating the analysis to the microstructure, although the contamination mark is usually considerably larger than the electron probe.

7.9.6 Wavelength Shift

Wavelength shift and changes of line profile with change in chemical combination are generally negligible, but can lead to errors when analysing the lighter elements. The magnitude of this effect depends on the resolution of the spectrometer as well as the shift in wavelength of the spectral line. For a spectrometer with poor resolution (e.g. the EDX) any change in line position is masked by the instrumental breadth of the line and can generally be ignored. Using a WDX the resolution is generally such that for elements of atomic number 14 (silicon) and less, allowance may need to be made for this effect. In principle, measurement of the integrated intensity is the correct procedure, but in using a WDX this is not practicable, and re-adjustment of the spectrometer to the peak intensity, or prior measurement of the peak shape and of the peak shift between specimen and standard and use of a correction factor to allow for the loss of intensity, have both been used successfully. Perhaps the best procedure, if it is practicable, is to use standards having a similar chemical structure as the specimen, so that no wavelength shift occurs.

7.9.7 Drift

Drift is a term usually reserved for a slow and hopefully steady change of probe current with time. This is usually due to movement of the filament in the electron gun, and should not be due to any other cause. Emission from the gun is most sensitive to filament position when the filament-grid spacing is small. In practice the source of any observed drift should be located—for example by using the filament centring adjustment to check whether filament movement was responsible for the decrease in probe current—and then considering what action can be taken to minimize the effect. Allowance can be made for a small steady change in probe current, if frequent measurements are made using a Faraday cage, but anything worse than 2 % per hour can generally be overcome by attention to the cause.

7.9.8 Incorrect Value of Accelerating Voltage E_0

This is generally only important if for some reason small values of overvoltage U_0 are used. In this case inaccuracy in the meter reading may cause error, and calibration of the voltage measuring circuit is necessary. This is best accomplished using a chain of standard high resistances and a potentiometer or a calibrated microammeter. The same end can be achieved by choosing an element which has an excitation voltage E_K in the range required, and then plotting the intensity of characteristic K radiation versus E_0 as measured on the instrument's meter. This characteristic intensity drops to zero when $E_0 = E_K$. The measurement can be repeated using other elements if necessary.

There is a further reason why the meter reading, even if the meter is accurate, does not give the correct value for E_0. This is due to the voltage fall across the bias resistor $i_B R_B$, where i_B is the beam current (leaving the electron gun) and R_B the value of bias resistor in use. This only becomes significant at low values of E_0.

7.9.9 Unexplained Systematic Errors

These arise with any instrument from time to time. Usually, if they lead to a serious error, their presence is detected, when after investigation the cause is located and either removed (by an appropriate repair) or a procedure is evolved which eliminates the effect of such error. To detect such errors it is good practice to include in the measurements regular analysis of a known specimen which is as similar as possible to the specimens being studied. This 'monitor' specimen is measured in terms of the standards that are used, and since the intensity ratios to be expected are known, any change, either in the instrument or the way it is being used, can be detected at the earliest opportunity and action taken to overcome the problem.

REFERENCES

1. P. M. Martin and D. M. Poole, *Met. Rev.*, 1971, **15**(150), 19.
2. T. O. Ziebold and R. E. Ogilvie, *Anal. Chem.*, 1964, **36**, 322.
3. T. R. Sweatman and J. V. P. Long, *J. Petrol.*, 1969, **10**, 332.
4. D. R. Beaman and J. A. Isasi, ASTM–STP 506, 1972.
5. H. E. Bishop, *X-ray Optics and Microanalysis, Fourth Int. Congress*, Orsay, 1966, p. 153 (eds. Castaing *et al.*). Hermann, Paris.
6. G. Springer, *Fortschr. Mineral.*, 1967, **45**, 103.
7. H. A. Bethe, M. E. Rose and L. P. Smith, *Proc. Am. Phil. Soc.*, 1938, **78**, 573.
8. A. J. Nelms, NBS Circular 577, 1956.
9. M. J. Berger and S. M. Seltzer, Nat. Res. Council Publication 1133, p. 205, 1962.
10. P. Duncomb and S. J. B. Reed, NBS Special Publication 298, p. 13, 1968.
11. J. Philibert and R. Tixier, NBS Special Publication 298, p. 13, 1968.
12. J. Philibert, *X-ray Optics and Microanalysis, Third Int. Symp.*, Stanford, 1963, p. 379 (eds. Pattee *et al.*). Academic Press, New York.
13. H. E. Bishop, *J. Phys. D: Appl. Phys.*, 1974, **7**, 2009.

14. K. F. J. Heinrich, in *The Electron Microprobe* (eds. McKinley *et al.*) p. 296. Wiley, New York, 1963.
15. B. L. Henke, R. L. Elgin, R. E. Leut and R. B. Ledingham, *Norelco Reporter*, 1967, **14**, 112.
16. J. Ruste and M. Gantois, *J. Phys. D: Appl. Phys.*, 1975, **8**, 872.
17. G. Love, M. G. C. Cox and V. D. Scott, *J. Phys. D: Appl. Phys.*, 1974, **7**, 2131
18. M. Green and V. E. Cosslett, *Proc. Phys. Soc.*, 1961, **78**, 1206.
19. E. H. S. Burhop and W. N. Asaad, *Adv. Atom. Mol. Phys.*, 1972, **8**, 163.
20. COR 2: J. Henoc, K. F. J. Heinrich and R. L. Myklehurst, NBS Technical Note 769, 1973.
21. EMPADR VII: J. Rucklidge and E. L. Gasparrini, Geology Department Report, University of Toronto, Canada, 1969.
22. FRAME: H. Yakowitz, R. L. Myklehurst and K. F. J. Heinrich, NBS Technical Note 796, 1973.
23. R. Woldseth, *X-ray Energy Spectrometry*, Kevex Corp., Calif., USA, 1973.
24. P. J. Statham, *X-ray Spectrometry*, 1976, **5**, 16.
25. G. Love, M. G. C. Cox and V. D. Scott, *J. Phys D: Appl. Phys.*, 1976, **9**, 7.
26. L. J. Gray and C. A. Wert, *Adv. X-ray Anal.*, 1969, **12**, 563.
27. B. L. Henke and R. L. Elgin, *Adv. X-ray Anal.*, 1976, **13**, 639.

8

Applications of Electron Probe Microanalysis

J. A. BELK

The Royal Military College of Science, Shrivenham, UK

8.1 INTRODUCTION

A number of articles have been written describing the various fields in which electron probe microanalysis has been applied.[1,2] The areas of application vary widely and there are few clear links between them. In this chapter three specific areas of application will be described. These are linked firstly by the fact that they are fairly new areas, and secondly because the success of the measurements depends on a knowledge of the distribution of ionization with depth in the sample. The three areas are the composition of small particles, the composition of thin layers and quantitative analysis for the light elements ($Z < 11$). A number of papers will be discussed that indicate various ways of obtaining useful experimental results in these areas of electron probe microanalysis, despite the fact that there is no universal agreement on the best mathematical model to describe ionizations with depth in the specimen.

8.2 PARTICLES AND SEGREGATION

The analysis of particles set in a matrix has always been one of the major applications of electron probe microanalysis. Clearly the minimum size of particle that can be analysed is of interest and various estimates have been made of the size of a component of a microstructure that will allow a satisfactory analysis. Reed[3] considered lamellae, cylinders and hemispheres. Three approaches are possible to analyse particles that are not large enough to contain the whole of the X-ray source. One can correct the results for the effects of the matrix, or extract the particles and study them in isolation, or make up special materials based on the measured composition and compare the properties of these with those under study. Examples will be given of all three approaches, none of which is of universal application, as often only one of them is possible in a given case. Equally, they all suffer from the criticism that one is either making assumptions without clear proof of their validity, or that one is making a new specimen whose composition can be measured to replace the original specimen whose composition cannot be measured.

Ryder and Jackel[4] studied the composition of mixed iron manganese sulphide inclusions in steel. The particles were from 2 to 10 μm in dimension and as their apparent sulphur content decreased it was found that their apparent iron content increased. This was because the X-ray source size was larger than the particle and some of the iron X-rays were originating in the matrix. The assumption was made that the (MnFe)S particles were stoichiometric, and that the X-ray source consisted of a weight fraction α of the mixed sulphide and a weight fraction $(1 - \alpha)$ of the matrix. After correcting for X-ray absorption and atomic number effects it is possible to find α for any given particle and to correct for the effect of the matrix. This raised the mean manganese content from 44·1 wt % to 54·2 wt %, giving a mean iron content of about 9 %. However, there was a correlation between the corrected manganese contents and the correction factor α, indicating a systematic error in the correction procedure. When the corrected manganese and iron values were plotted against α and the values extrapolated at $\alpha = 1$ an iron composition of 1·4 % was given. This is equivalent to extrapolating the measurements to large particle sizes, but was shown to be more economical and accurate than plotting the composition versus particle size. Even particles as large as 12 μm only gave values of α of 0·95, indicating that very large particles are required to contain the X-ray source.

The reason for the systematic error is that the X-ray source size is different for the three elements Fe, Mn and S because of their different X-ray critical excitation potentials, E_c. Since Fe and Mn have similar atomic numbers one can assume that they are both governed by the same correction factor. However, S is a much lighter element with a lower E_c and a different correction factor β must be used in this case, as the X-ray source size is larger. Again it is possible to find α and the corrected values of Fe do not correlate with the α value used, showing that there is no longer any systematic error. The mean iron content found by this improved technique was 1·6 % in good agreement with the previously extrapolated value. The whole of this technique for allowing for matrix effects in the analysis of small particles depends on the assumption of stoichiometry. Ryder and Jackel comment that if a fourth element Ti is present, whose effect on the MnS lattice is not known, one cannot assume a stoichiometric composition, hence the above correction procedure for matrix effects cannot be used. It is still possible to obtain a value for the various elements but it is much more difficult to show that the correction method used is a satisfactory one.

A case where it is not possible to apply a simple correction for the matrix effect on the composition of small particles was recently studied by Belk.[5] In this case it was required to determine the composition of a third phase in a quaternary alloy which occurred on the interface between the other two phases. One of the matrix phases was aluminium rich, the other zinc rich and the third phase contained more Mg and Cu than either of the other phases. There is a similar problem in X-ray source size as that of Ryder and Jackel, in that Cu and Zn have high values of E_c hence a relatively small source size, and Al and Mg have low values of E_c and so relatively large X-ray source size. Indeed Reed's[3] formula gave the minimum thickness of lamellae suitable for quantitative analysis of Zn and Cu, at the 12 kV accelerating voltage chosen for this study, as being about half that required for Mg and Al. As the specimens were not lamellar, the exact values are of no great significance, but the relative values are important in illustrating the difference in X-ray source size for various elements. Thus not only would one have needed the α and β values of Ryder and Jackel but a third factor indicating what proportion of the two matrix phases was responsible for the matrix effect. As this more complex problem is not soluble algebraically, the approach adopted was to make the best estimate of the composition of the third phase, and then to make up an alloy of that composition and study it.

Analysis of the 2 to 5 μm diameter particles showed the following composition after correcting for X-ray absorption and atomic number

$$\begin{array}{lll} \text{Mg} & 4 \cdot 90 \text{ wt} \% & (0 \cdot 124) \\ \text{Cu} & 5 \cdot 38 \text{ wt} \% & (0 \cdot 117) \\ \text{Al} & 6 \cdot 77 \text{ wt} \% & (0 \cdot 381) \\ \text{Balance Zn} \end{array}$$

The figure given in brackets in each case is the coefficient of variation, i.e. the standard deviation of all the measurements divided by the mean value. This is of use in assessing the relative spread of analyses for each of the elements. Clearly Al gives the greatest spread due to the large X-ray source size and the high concentration of Al in one of the matrix phases. A special alloy was made up close to this composition and after soaking for 300 h at 360 °C was found to be two phase. The matrix was the phase under study while the particles within it were of the high temperature zinc rich phase α'. The analysis of the matrix was found to be

$$
\begin{array}{lll}
\text{Mg} & 6.59 \text{ wt}\% & (0.009) \\
\text{Cu} & 8.66 \text{ wt}\% & (0.106) \\
\text{Al} & 2.63 \text{ wt}\% & (0.042) \\
\text{Balance Zn} & &
\end{array}
$$

The obvious difference from the previous set of results is the reduced value of the Al composition and the very much reduced scatter on the Al results. Both are the result of removing the effect of the two phase matrix in which the small particles were set. There is still a fairly high scatter on the copper results, but this is because the accelerating voltage of 12 kV is quite close to the critical excitation potential for copper of 8.05 kV, and with a fairly low concentration the accuracy suffers. One further advantage of making up the special alloy is that one can undertake X-ray diffraction experiments to determine the crystal structure which is also difficult on very small particles. In this case the third phase was found to be isotypic with Mg_2Zn_{11} and $Mg_2Cu_6Al_5$ and its analysed composition is very close to seven parts of Mg_2Zn_{11} and one part of $Mg_2Cu_6Al_5$. The important metallurgical conclusion from this work is that although 0.1 wt % Mg is soluble in the high temperature α' phase, it is not soluble in either the α or β low temperature phases as shown in Fig. 8.1. The magnesium is thus required to separate in the form of the third phase studied, and its presence influences the kinetics of transformation and the mechanical properties of the transformed material.

An example of the necessity of removing very small particles from their matrix in order to study their composition and structure is the work of Jacobs et al.[6] on the nucleation of graphite nodules in chill cast iron. Particles were found of ∼ 1 μm in diameter at the centre of graphite nodules and within the iron matrix. In order to study the composition and structure of such small particles it was necessary to remove them from the iron matrix. The instrument used was EMMA-3,[7] a combined electron microscope/microanalyser, which produces a 0.1-μm-diameter electron probe for analysis. The particles were about 0.3-μm-thick hexagonal

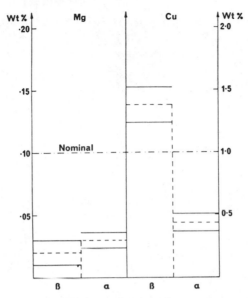

Fig. 8.1. Measured concentrations of magnesium and copper in Zn–Al alloy phases α and β compared with the nominal overall composition in the alloy.[5]

platelets. When the probe was placed on the centre of the particle and then on the outer region the following differences were found:

1. Mg, Al and Si were found throughout, but the central region was enriched in Mg but depleted in Si relative to the outer region.
2. A trace of S was present in the outer region.
3. The particle contained oxygen.
4. Some particles contained a trace of Ti or Fe in their outer region. The overall composition of the particles depended on the alloy from which they were removed.

In order to clarify the structure of the particles, electron diffraction studies were used to supplement the microanalysis data. The final picture of the particle structure that emerges is that the centre is of a CaMg sulphide and the outer shell is a Mg, Al, Si, Ti oxide with a spinel structure. It is essential to remove such small particles from their matrix in order to obtain information about their composition, as the particle would only constitute a small part of the X-ray source in a solid material.

In studying more gradual changes in composition such as one finds in segregation in a single phase material, one can either use the technique of a

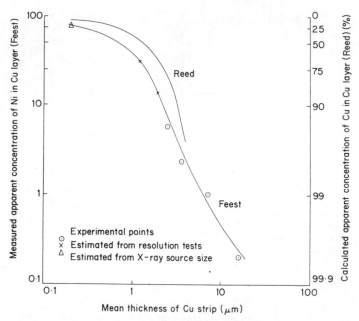

Fig. 8.2.   Comparison of Reed's[3] and Feest's[8] results on apparent composition of thin copper layers in nickel.

line scan or of a series of random point analyses. Feest[8] shows that there are many advantages in choosing the random point analyses. If the measurements are ranked in increasing order of concentration this allows a more accurate assessment of the maximum and minimum concentrations to be obtained. The point counting also has the advantage that the sensitivity of analysis is increased because more time is spent on each individual spot than in a line scan. He also studied the spatial resolution of microanalysis by looking at laminated Cu/Ni specimens. His results make an interesting comparison with those of Reed[3] already referred to. Feest plots the measured apparent concentration of Ni in thin Cu layers surrounded by Ni, whereas Reed plots the calculated apparent concentration of Cu in Cu layers—thus Feest's 10% is equivalent to Reed's 90%. When their results are compared we find the result given in Fig. 8.2. Unfortunately Feest's experimental points only just overlap with the range of thickness covered by Reed, but where they do overlap the agreement is satisfactory. In the lower thickness range Feest estimated the compositions from resolution experiments and from the X-ray source size. These estimated values differ from the calculations of Reed rather more than Feest's experimental measurements of composition do. It is interesting to note that the thickness of layer required to give 99 % Cu or alternatively 1 % Ni is given by both methods as about 6 μm. Accurate analysis of particles would require them to be considerably bigger than this in linear dimension.

8.3 COMPOSITION OF THIN LAYERS

Estimation of the thickness and composition of thin surface layers has been of interest since the early days of electron probe microanalysis. Cockett and Davis[9] used the experimental results of Castaing and Deschamps[10] on the X-ray intensity generated by an electron beam as a function of depth to calculate thickness calibration curves. Measurements made on the X-ray emission from the film or its substrate could be used, and an agreement with other thickness measurements of about 15 % was claimed. Analysis was limited to 29 kV as this was the only accelerating voltage for which measurements of X-ray intensity with depth had been made.

A similar approach has recently been made by Bishop and Poole[11] but with two modifications. Firstly they used the Monte Carlo calculations of Bishop[12] to describe the electron penetration and energy loss in the specimen rather than experimental measurements. They admit that the calculation is based on simplified scattering cross sections and that a more refined scattering model would give better results for thicker films. Secondly they show that if all energies are expressed as a fraction of the incident energy, and all distances as a fraction of the Bethe range, it is possible to produce a set of curves plotting $p(t)$ the integral distribution of ionization with depth against both overvoltage ratio and fraction of the Bethe range for a series of six elements. In the simplest case where absorption corrections can be neglected and the substrate and film have similar atomic numbers, the measured intensity ratio from the film k_A is equal to the integral distribution $p(t)$ and the curves allow the value of t to be determined.

The Bethe range, which is the total length of the ionizing electrons' path in the target, is also plotted for various beam energies and atomic numbers of the target. The mass thicknesses covered are from less than 1 μg/cm^2 to several mg/cm^2 whereas Cockett and Davis could only go to 1·5 mg/cm^2 due to the limitation of working at 29 kV.

If absorption cannot be neglected Bishop and Poole's simplest equation becomes

$$k_A = C_A p(t) \exp\left(-\frac{\chi_A t}{2}\right)$$

The exponential term is an approximate absorption correction; more accurate forms are also discussed.

As a comparison of the two methods described, Bishop and Poole's method was used to construct the intensity ratio versus thickness for chromium coatings and this was compared with the values given by Cockett and Davis. Bishop and Poole gave slightly smaller thicknesses, but they admit that on average, thicknesses deduced by their method were 15% lower than those measured by chemical techniques. The discrepancy between the two methods was just greater than 15% but as Cockett and Davis only claim 15% absolute accuracy it is probable that both claims are justified. Clearly Bishop and Poole's is much the more versatile method and yet is quite simple to calculate by hand, and until a more refined scattering model is available will give quick quantitative information in a wide variety of situations.

An alternative method has been proposed by Colby[13] based on the Philibert theory of analysis of bulk specimens. He used the Philibert expressions for X-ray emission which require a knowledge of the mean electron energy on going from the thin layer to the substrate and the number and energy of electrons backscattered by the substrate into the thin layer. The expressions must be integrated by computer and they allow the thickness and composition of an alloy layer to be determined if one characteristic line is measured for each constituent element. A review of the use of electron probe microanalysis on thin specimens is given by Hall.[14]

The limitation on measurement of thin layers on a solid substrate is due to the contribution of the substrate in adding to the background radiation and in backscattering electrons that have already passed through the layer. The limitation is substantially removed if the substrate can be removed. The previous methods mentioned can measure films of metals just less than 1 nm thick or about 10^{16} atoms/cm^2. It is possible to measure films of only 10^{12} atoms/cm^2 or 10^{-3} monolayers if they are deposited on carbon films some 20 nm thick. Ecker[15] claims a sensitivity for the electron probe microanalysis of such films of 10^{12} atoms/cm^2 which makes the method competitive with Auger electron analysis, low energy ion scattering and autoradiography. The intensity of a characteristic X-ray peak is directly proportional to the thickness of the film from which it originates. A rough calibration of spectrometer efficiency enabled the experimental relative intensities of X-rays emitted from 10^{15} atoms/cm^2 to be calculated and compared with the product of the theoretical ionization cross section and the fluorescent yield (see Chapter 7). Agreement was within a factor of two which was considered satisfactory when the experimental difficulties and theoretical uncertainties were taken into account. Only a high efficiency spectrometer such as that of EMMA was able to measure 10^{-3} monolayers. A scanning electron microscope with energy dispersive spectrometer was only capable of measuring down to 10^{-1} monolayers.

There are occasions when the composition of a thin film is required rather than its thickness. Mitra and Hall[16] have developed a network method based on energy dispersive analysis which allows this to be obtained. As long as the specimen and standards are thin enough to transmit the electrons with only a small energy loss, the ratio of the characteristic X-ray intensity to the continuum intensity for both specimen and standard can be used as a measure of composition.[17] Analysis of iron–nickel alloy films of about 25 nm thickness gave compositions close to those of the bulk material and showed that the analysis was independent of the thickness of the films.

Lorimer[18] has reported a very simple way of obtaining microanalysis results from thin specimens. If the specimen is thin enough to transmit 100 keV electrons and is examined with the energy dispersive analysis equipment in EMMA, the relative intensity of characteristic X-rays from two elements will depend only on their relative stopping power and weight fraction ratio. A series of scaling factors k can be determined from calibration of the instrument such that

$$k_{AB} = \frac{C_A}{C_B} \times \frac{I_B}{I_A}$$

Thus the relative intensities of peaks in the X-ray spectrum I can be transformed into relative weight fractions C by means of the factors k. The only limitations are that all the elements present must be detected and that a new calibration will be required if the window is changed or a different accelerating voltage is used. This technique is discussed in more detail in Chapter 3.

8.4 QUANTITATIVE LIGHT ELEMENT ANALYSIS

The development of the technique of electron probe microanalysis has seen a lowering of the minimum atomic number element that could be satisfactorily analysed. In the early 1960s the group of elements with atomic number of 9 or less were designated light elements and they all presented particular problems to the microanalyst. Firstly, their X-ray wavelengths were longer than could conveniently be diffracted by available natural crystals, secondly, these soft X-rays were more difficult to detect, thirdly, there was no fully satisfactory correction procedure and fourthly, the mass absorption coefficients were either unknown or very inaccurate. A great deal of work on all these aspects has considerably improved the situation but this group of light elements still pose problems in comparison with the heavier elements.[19] The techniques of diffraction and detection of soft X-rays are dealt with in Chapter 6 and the work on correction procedures and the constants used is covered in Chapter 7. In this section some of the practical aspects of light element analysis will be discussed in the hope of illustrating its possibilities in a given experimental situation.

It should be stated that apart from conventional electron probe microanalysers it is also possible to detect the presence of light elements in transmission electron microscope specimens by measuring the K energy loss of the electrons.[20] This is claimed to be advantageous for the very light elements which show an increase in energy loss signal but a decrease in X-ray emission. Although corrections for absorption and fluorescence are not required, the instrumental aspects are quite complex, but do not allow quantitative information to be obtained. Spatial distribution information can be obtained by forming an image with electrons that have suffered a particular K energy loss. In this way the distribution of BN, MgO and AlN particles down to 30 nm in size has been illustrated on carbon support films.

Many factors influence the quantitative accuracy of light element microanalysis results which are not important in heavier element microanalysis. For example, the preparation of the specimen and the necessity to make it conducting by evaporating a surface layer of a material that will not interfere with the soft X-ray emission. Also there is the problem of producing suitable standards of elements that are gaseous or non-conducting solids in their pure form. There are the difficulties that have already been mentioned in Chapter 7 of change of X-ray peak shape and position with a change of chemical bonding. Another difficulty is the contamination which is caused on the specimen when the hydrocarbons present in the vacuum system are broken down to carbon by the action of the probe on the surface of the specimen. These are all problems that must be met and dealt with as part of the standard light element microanalysis technique before satisfactory raw analysis figures can be guaranteed. The rest of this section will be concerned with a few examples of practical light element microanalysis to indicate its scope and application.

The determination of carbon profiles in steel after carburization is an obvious application. There is a very large body of empirical 'know how' but very few scientific analyses of the process and its results. Two papers, by Swaroop[21] and Duerr and Ogilvie,[22] have studied this process and it is interesting to compare their results. They both consider the contamination that occurs in microanalysis resulting in increasing carbon count rate with counting time. Both consider that a fine gas jet impinging on the specimen is the most satisfactory way of preventing the build-up of a carbon contamination spot under the electron probe. Swaroop used helium and found that the device was only satisfactory when used in conjunction with a cold finger. Duerr and Ogilvie used dry filtered air and found this to be the best. The limit of detectability is calculated differently in the two papers. Swaroop uses a simple formula and arrives at a figure of 0·03 wt %C. Duerr and Ogilvie use a more sophisticated formula and quote a value of 0·1 wt %C. It should be mentioned however that the latter workers only used count rates a quarter of that of Swaroop for a given carbon content. This may well be due to their choice of probe current but it will lead to a difference factor of two in the limit of detectability in the two cases. In order to achieve these values, counting times were between 200 and 500 sec for a given composition, and Duerr and Ogilvie state that a whole experimental run took from 12 to 18 h to complete. Both groups used alloy standards rather than any of the pure forms of carbon. This minimizes the difficulties in choice of standard, correction procedure and constants and Duerr and Ogilvie state 'A calibration curve is really necessary in order to have any confidence in the experimental data for the light elements.' Martensitic steel standards were used in both cases.

The variation of carbon with depth obtained in the two cases is shown in Fig. 8.3. Duerr and Ogilvie used an 871 H steel carburized for 6 h at 927 °C, quenched and tempered. Swaroop used a 9310 steel gear tooth but gives no details of the carburizing treatment. The difference in the shape of the curves is very apparent and neither look very close to the error function curve predicted by Fick's second law of diffusion. The diffusion coefficient of carbon in steel is known to vary with the carbon content and this may account for the discrepancy from the ideal form. The difference in shape of the two curves is just greater than the quoted experimental error, and is probably due to the much higher alloy content (3·2 % Ni 1·2 % Cr) of the 9310 steel than that of the 8717 steel (0·55 % Ni 0·5 % Cr). Swaroop compared his carbon determination with the hardness variation of the steel with depth. He found hardness to be rather insensitive at the higher carbon contents and not to vary at all below 0·2 %C. In fact his hardness curve had rather the shape of the carbon penetration curve of Duerr and Ogilvie. The differences shown up in these two papers are clearly worth further study to establish the validity of the carbon determinations, and the effect of alloy composition and carburizing variables on the diffusion of carbon into steel.

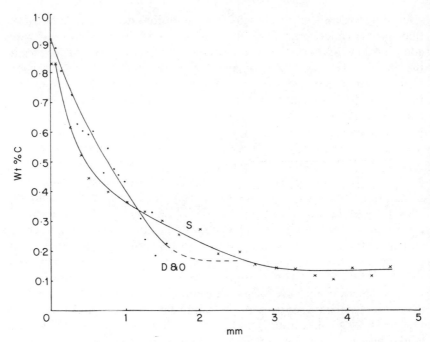

Fig. 8.3.   Comparison of results of Swaroop (S)[21] and Duerr and Ogilvie (D&0)[22] carbon concentration with depth below surface of carburized steel.

The difficulties over choice of correction procedure and constants can be overcome in the case of identification of compounds rather than solid solutions as the X-ray peak shifts in accordance with the valence state and bonding in the compound. This is a particularly useful relationship when studying corrosion products which may be thin and mechanically weak and attached to a metal substrate. Such a specimen would be unsuitable for quantitative microanalysis but if the precise wavelength of its X-ray peaks can be measured these will reveal the composition of the compounds. Workers at Pennsylvania State University have shown that peak shift is directly related to the cation valence state.[23,24] The oxygen Kα peak moves progressively as the molar ratio of metal to oxygen is increased. A wavelength change of 0·001 nm to 0·005 nm corresponds to a change in molar ratio of from 0·5 to 1·0 in the case of titanium oxides. With the oxygen peak occurring at 2·36 nm this represents a change in wavelength of only 0·2 % at the maximum but the resolution of the clinochlore crystal allows this, and the reproducibility of the peak settings was claimed as

$\pm 0.000\,27$ nm enabling all the measurements to be made satisfactorily. The main limitation is that the electron probe must be spread over a fairly large spot in order to limit the decomposition of the oxides which the beam energy induces. The measurement of peak shifts not only for the light elements but also for soft X-rays from the heavier elements should lead to a better understanding of the electronic structure of compounds and also a much simpler method of microanalysis.

REFERENCES

1. W. J. M. Salter, *Micron* (London), 1973, **4**, 307–31.
2. J. I. Goldstein, *Electron Probe Microanalysis* (eds. A. J. Tousimic and L. Marton), pp. 245–90. Academic Press, New York, 1969.
3. S. J. B. Reed, *Proc. Fourth Int. Conf. X-ray Optics and Microanalysis* (eds. R. Castaing, P. Deschamps and J. Philibert), pp. 339–49. Hermann, Paris, 1966.
4. P. L. Ryder and G. Jackel, *Zeit. für Metallk.*, 1972, **63**, 187–92.
5. J. A. Belk, *Micron* (London), 1974, **5**, 201–7.
6. M. H. Jacobs, T. J. Law, D. A. Melford and M. J. Stowell, *Metals Tech.*, **1**(11), 1974, 490–500.
7. M. H. Jacobs, *J. Microsc.*, 1973, **99**(2), 165.
8. E. A. Feest, *J. Inst. Metals*, 1973, **101**, 146–9.
9. G. H. Cockett and C. D. Davis, *Brit. J. Appl. Phys.*, 1963, **14**, 813–16.
10. R. Castaing and P. Deschamps, *C.R. Acad. Sci. Paris*, 1953, **237**, 1220.
11. H. E. Bishop and D. M. Poole, *J. Phys. D.*, 1973, **6**, 1142–58.
12. H. E. Bishop, *J. Phys. D.*, 1974, **7**, 2009–20.
13. J. W. Colby, *Advances in X-ray Analysis* II (eds. J. B. Newkirk, G. R. Mallett and H. G. Pfeiffer), pp. 287–305. Plenum, New York.
14. T. A. Hall, *Advances in Analysis of Microstructural Features by Electron Beam Techniques*, pp. 121–39. Metals Soc., 1974.
15. K. H. Ecker, *J. Phys. D.*, 1973, **6**, 2150–6.
16. S. K. Mitra and T. A. Hall, *J. Phys. D.*, 1972, **5**, 1502–12.
17. D. J. Marshall, Ph.D. Thesis, University of Cambridge, 1967.
18. G. W. Lorimer, *ibid.* ref. 14, p. 140.
19. V. D. Scott, *ibid.* ref. 14, pp. 141–92.
20. R. F. Egerton, *ibid.* ref. 14, pp. 67–71.
21. B. Swaroop, *Mater. Eval.*, 1973, **31**, 185–7.
22. J. S. Duerr and R. E. Ogilvie, *Anal. Chem.*, 1972, **44**, 2361–7.
23. P. D. Gigl, G. A. Savanick and E. W. White, *J. Electrochem. Soc.*, 1970, **117**, 15–17.
24. H. B. Kranse, G. A. Savanick and E. W. White, *J. Electrochem. Soc.*, 1970, **117**, 557–8.

9

Kossel X-ray Microdiffraction

N. SWINDELLS

University of Liverpool, UK

9.1 INTRODUCTION

The electron beam in scanning electron optical instruments can be at a high enough energy (> 15 keV) to generate characteristic X-rays from crystalline metallic or mineralogical specimens. The penetration of the electrons therefore generates what is effectively a point source of these X-rays inside a single crystal (of size > 2 μm) when the electron beam is stationary.[1] If the Bragg conditions can be satisfied from the wavelengths present then the completely divergent beam produces an X-ray diffraction pattern called after Kossel.[2,3] An example is shown in Fig. 9.1. The relationship which this method has with more conventional X-ray diffraction techniques is shown in Table 9.1. All values of the Bragg angle are present and therefore all planes capable of reflecting do so. The Kossel pattern therefore contains

TABLE 9.1
A Comparison of X-ray Diffraction Methods

Specimen	X-ray wavelength	Direction of incident beam	Diffraction effect	Method
Single crystal	Continuously variable	Fixed	Spots	Laue
Single crystal	Fixed	Restricted	Spots	Rotating crystal etc.
Polycrystal	Fixed	Continuously variable	Lines	Powder
Single crystal	Fixed	Continuously variable	Lines	Kossel

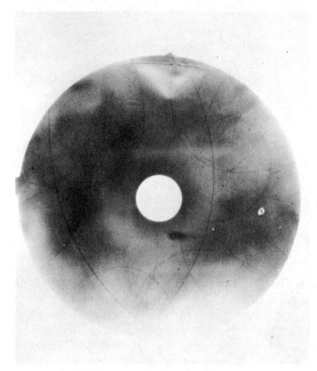

Fig. 9.1.   Back reflection Kossel pattern from a grain of austenite (original
contrast).

plane spacing information and, because it is generated in a single stationary
crystal, orientation information as well. The pattern can be recorded on
suitable photographic emulsion as a reflection pattern, with the lines
showing as mostly dark against a lighter background, or as an absorption
or deficiency pattern after transmission through a suitably thin specimen.

The main advantage to be gained from generating these patterns with
scanning electron beam instruments is that the beam can be located
accurately within the crystal of interest. The pattern comes from a region of
approximately $2\,\mu m$ in diameter centred on the source. Therefore,
individual phases in a multiphase mixture can be examined by X-ray
diffraction on the same scale as one would use for their examination by
optical microscopy and on the same kind of specimens.

The pattern is related absolutely to the crystal because it is not processed
by the instrument and it is independent of the direction of the incident

electron beam. The Kossel pattern can therefore provide accurate orientation data which can be combined with two-surface analysis to provide full orientation information about features which can be observed by optical microscopy and low magnification scanning microscopy.

The following sections provide some practical guidance on the methods of obtaining good Kossel patterns and a summary of methods used to interpret them.

9.2 THE GEOMETRY OF KOSSEL PATTERNS

Figure 9.2 seeks to show the main features of the diffraction effect by means of the Ewald construction. The completely divergent source allows an infinitely variable angle of incidence in a fixed lattice. The effect is as though the Ewald sphere rolls around inside the limiting sphere without restriction. Diffraction from two planes is shown with incident rays I_1 and I_2 respectively. The diffracted rays R_1 or R_2 will form a cone by rotation of R_1 or R_2 about the reciprocal lattice vector corresponding to the reciprocal lattice point for each plane, RLP1 or RLP2. These cones have a half angle of the complement of the Bragg angle for the plane.

There is a logical difficulty in using this model because with a fixed lattice

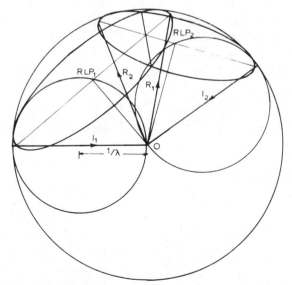

Fig. 9.2. Kossel X-ray diffraction and the Ewald construction.

I_1 and I_2 should come from the same point, and hence there would be a different position of the *limiting sphere* for each incident ray, but this would also need a different origin for the reciprocal lattice for each reflection which destroys the concept of the Ewald construction. Despite these problems the pattern behaves as though Fig. 9.2 were the correct description of the diffraction effect and hence it is very profitable to continue to apply it.

The two reflection cones are shown to intersect in real space, and these intersections are an important feature of the pattern and can be used directly in its interpretation. The intersection can be treated either as an intersection between the cone surfaces or as an intersection between the planes circumscribed by the cones on the limiting sphere. The plane section formed in this way is sometimes called the Kossel plane. If the Kossel planes meet in a line then two intersections are formed, but three planes can meet in a point and there are three types of intersection:

1. *Lens intersections* (Fig. 9.3(a)) are formed from two cones when $(\alpha_1 + \alpha_2) > \phi_{12}$ where α_1 and α_2 are the half angles of the cones and ϕ_{12} is the angle between their axes. The existence of such an intersection is dependent on the wavelengths present and the lattice dimensions, and these intersections have been used as the basis for the high-precision lattice parameter determinations made by Ogilvie's group.[4,5]

2. *Persistent intersections* (Fig. 9.3(b)). When the reflecting planes are tautozonal the cones form a persistent pair of intersections whose existence is independent of the wavelength. Several cones may intersect at two common points in this way and the conditions for this to occur have been investigated by several authors.[6-10] Mackay[8] and Frazer and Arrhenius[9] showed that the necessary condition for three planes is that the determinant:

$$\begin{vmatrix} h_1 & k_1 & l_1 \\ h_2 & k_2 & l_2 \\ h_3 & k_3 & l_3 \end{vmatrix} = 0$$

where $(h_1 k_1 l_1)$ are the Miller indices of the reflecting planes. Yakowitz[10] showed that a sufficient condition is that the four determinants:

$$\begin{vmatrix} h_1 & k_1 & l_1 \\ h_2 & k_2 & l_2 \\ h_3 & k_3 & l_3 \end{vmatrix} = \begin{vmatrix} h_1 & k_1 & Q_1 \\ h_2 & k_2 & Q_2 \\ h_3 & k_3 & Q_3 \end{vmatrix} = \begin{vmatrix} h_1 & Q_1 & l_1 \\ h_2 & Q_2 & l_2 \\ h_3 & Q_3 & l_3 \end{vmatrix} = \begin{vmatrix} Q_1 & k_1 & l_1 \\ Q_2 & k_2 & l_2 \\ Q_3 & k_3 & l_3 \end{vmatrix} = 0$$

where $Q_i = \lambda/2d_i$ ($i = 1$, 2 and 3).

3. Single intersections (Fig. 9.3(c)) are formed when three Kossel planes

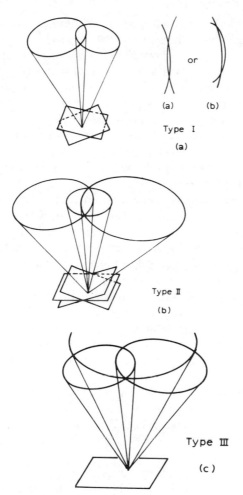

Fig. 9.3.   Types of intersections between Kossel cones. (a) Lens, (b) persistent, (c) single.

meet only at one point. The existence of an intersection of this form is dependent on the wavelength and the lattice constant and is such a unique combination of circumstances that it forms a powerful and rapid method of determining the lattice constant directly from the wavelength and the indices of the reflection only.[9]

The geometry of the recorded pattern depends on the projection of the cones on to the film. Cylindrical films have been used,[11] but flat films are easier to position and interpret, and the pattern is then a gnomonic projection with the centre of the projection at the point where a line from the source is a normal to the film. This point is also called the *pattern centre*.

The early investigators used spherical trigonometry[7,11] or a stereographic projection[7,12] as the basis for measurements and an independent knowledge of the pattern centre is essential for these methods. Spherical trigonometry is a cumbersome approach to the problem and vector methods are more generally useful. Vector methods were introduced nearly simultaneously by Bevis and Swindells,[13] Ryder et al.[14-16] and Morris.[17] Bevis and Swindells and Morris still used an independent determination of the pattern centre in their methods but the major step forward was made by Ryder et al. who showed that for orientation studies a knowledge of the pattern centre is not necessary in their method. Later Bevis et al.[18] suggested that the pattern centre could be calculated from the projected cones because the property of the gnomonic projection is that the major axes of the conic sections formed by the cones must extrapolate to the projection centre, which is the pattern centre. This suggestion was developed by Harris and Kirkham[19] into a method of determining plane spacing and orientation results from measurements made only on the pattern. The Harris and Kirkham analysis of the pattern along the lines suggested by Bevis has been developed further by Dingley.[20] Tixier and Waché[21] have extended the basic method of Ryder et al. to cover the determination of lattice parameters as well.

These developments in the methods of interpretation have had an important influence on the experimental method because, if it is not necessary to establish the pattern centre by some external reference, many of the restraints which existed on the design of the camera and the method of measurement are removed.

In the vector analysis methods the cones are assumed to have a common origin at the surface of the specimen. The axis of the cone is then described by the components of a vector referred either to an arbitrary set of axes (e.g. on the film), the crystal lattice base, or the reciprocal lattice base. The positions of the cone generator are describable in the same way. Cones will intersect along lines meeting at the origin and hence the lines will also be vectors. The components of the lines of intersection are also the Miller indices of the intersection when the base of the vectors is the crystal.[13] This property is useful because the position of a point of intersection on the film is easily measured and hence is a precise value to use in determining the orientation of the pattern, or the position of a cone generator.

Bevis and Swindells[13] combined the use of the direct and reciprocal

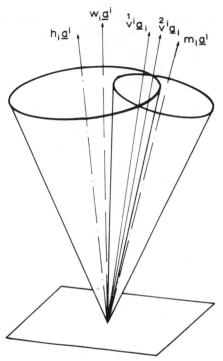

Fig. 9.4.   A vector description of the intersection of two cones: $h_i\mathbf{a}^i$ and $m_i\mathbf{a}^i$ are the cone axes and the normals to the planes with Miller indices h_i and m_i (i = 1 to 3), $w_i\mathbf{a}^i$ is the normal to the specimen surface, $V^{1i}\mathbf{a}_i$ and $V^{2i}\mathbf{a}_i$ are the lines of intersection of the two cones.

lattice bases so that directions were referred to the crystal lattice base (represented by the three vectors $\mathbf{a}_i$) and crystal planes were represented by their normals which were described by vectors referred to the reciprocal lattice base (represented by three vectors $\mathbf{a}^i$). The two bases are related by:

$$\mathbf{a}_i . \mathbf{a}^j = \delta^i_j \quad (\delta^i_j = 1 \text{ when } i = j \text{ and } 0 \text{ when } i \neq j)$$

A direction is therefore the vector V written as the contravariant vector[22] $V^i\mathbf{a}_i$ where the summation convention is used, so that:

$$V^i\mathbf{a}_i = \sum_{i=1}^{3} V^i\mathbf{a}_i$$

The intersection of two cones is shown in Fig. 9.4, and Bevis and Swindells showed that the Miller indices of the intersections are the two solutions V^{1i} and V^{2i} obtained by solving three equations for V^i:

$$\lambda/2 m_i m_j g^{ij} = m_i V^i \tag{1}$$

$$\lambda/2 h_i h_j g^{ij} = h_i V^i \tag{2}$$

$$V^i V^j = 1 \qquad (\text{i or j} = 1, 2 \text{ and } 3) \tag{3}$$

In these equations m_i and h_i are the Miller indices of the reflecting planes, $V^{1i}\mathbf{a}_i$ and $V^i\mathbf{a}_i$ are the lines along which they intersect and $g^{ij} = \mathbf{a}^i . \mathbf{a}^j$ is the metric tensor or metrical constant of the reciprocal lattice base.[23] Equations (1) and (2) are complete descriptions of the cones with axes $m_i\mathbf{a}^i$ and $h_i\mathbf{a}^i$ respectively. The indices of the intersections will be non-rational and it is usual to rationalize them so that

$$(V^1)^2 + (V^2)^2 + (V^3)^2 = 1$$

Notice that with the use of the metrical constant this analysis applies to any crystal base.

9.3   CAMERA DESIGN

The design of the camera and the methods of interpretation interact, as they do with all X-ray diffraction techniques. The main requirements are:

1.   The relationship between the pattern and the specimen must be accurately known, for orientation measurements.
2.   The emulsion should record only the X-rays produced and diffracted in the specimen and not electrons scattered from the specimen or the X-rays these electrons can produce.

An earlier requirement that the pattern centre be established by the camera design no longer exists with modern interpretation methods but the precision with which it can be located by these methods can be influenced by the camera design.

There have been three main classifications of camera design describable by the relation of the film plane to the specimen surface and the incident electron beam. All three of them are influenced strongly by the method of interpretation used by the designers. These three types are:

1.   Film and specimen surface parallel, normal to the electron beam and centred on it.[24,49]

2. Film and specimen surface parallel but both inclined to the electron beam.[25]

3. Specimen inclined to the electron beam with film and electron beam parallel.[26,27]

Important requirements in addition to the main ones listed above affect the interaction these designs have with the electron beam instrument used to generate the source. It is useful, for example, to intercept a solid angle larger than 60° because the easier visual interpretation this provides speeds up the detailed examination and the precision of the interpretation is improved. If the solid angle is greater than about 100° the distortion of the gnomonic projection becomes too great and the lines become broadened. For a given solid angle a large film is easier to investigate than a small one, although some investigators enlarge films as routine before measurement to improve the precision, so this requirement is the least serious. This is fortunate, because the main problems in the design are caused by trying to fit as large a film as possible to intercept as large a solid angle as possible in the vicinity of the specimen in a scanning electron microscope.

The approaches to the design problems represent two different philosophies. Some designers take the viewpoint that the application of Kossel microdiffraction would be to situations sufficiently different from those examined in normal scanning microscope operation to justify using the instrument only as a source of X-rays and replacing the usual stage by one specially designed to give the best conditions for recording diffraction patterns. This point of view has been criticized by Veith and Yakowitz[28] on the grounds that it uses an expensive and powerful instrument in a relatively simple way and restricts its normal use. Their solution was to use a simple electron beam instrument to generate the source in a purpose-built camera. The philosophy behind the design of Bevis and Swindells[26,27] represents a compromise. It is intended to fit permanently into the specimen chamber without causing a major alteration and without affecting the normal operation of the instrument so that patterns would always be available from suitable specimens without the need either to transfer them to another instrument or to exchange large parts of the instrument. It is the only design which allows reflection and transmission patterns to be recorded with the same film position but the criticism of Veith and Yakowitz still applies and the accuracy has not yet been established.

The requirement to record only the X-rays produced and diffracted in the specimen is not a trivial one. The precision of measurement of any X-ray diffraction pattern depends on the contrast of the lines. In other X-ray diffraction methods the width of the lines is an important additional effect but, since that is not a problem with Kossel lines, pattern contrast remains as the main limitation on interpretation. Since the intensity of the lines is fixed by the X-radiation and the reflecting plane, the contrast in the reflected pattern is controlled by the intensity between the lines. Any additional ionizing radiation will destroy the true contrast between the Bragg and non-Bragg scattered radiation from the specimen. There are considerable quantities of high-energy electrons scattering around in the specimen chamber of a scanning microscope. The greater proportion are those with the highest energies travelling back close to the direction of the incident beam. The energy distribution curve peaks sharply at $E/E_0 \simeq 0.8-0.85$[29] and the number of backscattered electrons varies linearly as $\cos \alpha$, where α is the angle between the ingoing and outgoing trajectories.[30] Films in all reflection positions have to be screened from these electrons, and domestic aluminium foil has been commonly used. Unfortunately the electrons have sufficient energy to excite X-rays from any metal foil in their path and since the thin foils easily transmit their own radiation (20 % for Al Kα generated in an 18 μm Al foil) there can still be unwanted blackening of the film. It is necessary to use thick screens or combination windows to trap the electrons and the X-rays they produce. Domestic aluminium foil (18 μm thick) coated with 1 μm of copper on the film emulsion side gives a transmission of <0.1 % for Al Kα radiation whereas the transmission of Cu Kα radiation is hardly affected. Thick plastic (e.g. Mylar 20 μm thick) in front of the aluminium foil is an alternative approach. Having trapped the direct radiation there can still be X-radiation as a result of electrons hitting surfaces in the microscope, lenses, stages, etc. The problem is worse than the similar and well-known problem with energy dispersive detectors because of the larger collecting area provided by the film.

9.4   EXPERIMENTAL METHODS FOR BACK REFLECTION PATTERNS

9.4.1 Specimen Preparation

Back reflection Kossel patterns from polycrystalline specimens obtained in electron probe instruments can be regarded as extending the information available with optical microscopy, and hence the preparation of the specimens will follow the requirements of that technique. There is, however, an additional requirement, in that as well as providing a surface with the necessary features clearly identifiable, the surface must be free from deformation. The effect of surface preparation is very noticeable. Figure 9.1 was obtained from austenite after a final mechanical polish for 20 min on Selvit cloth using $\frac{1}{4}$ μm diamond paste, whereas prior to this treatment the pattern was hardly discernible. The methods for attaining deformation-free surfaces with mechanical grinding and polishing have been described by Samuels[31] and it is clear from this work that while mechanical methods can produce a severely deformed surface they need not, and they are the preferred methods for obtaining a flat surface. Despite all the care which should be taken with mechanical methods, patterns can sometimes be improved by etching or by light electropolishing. Because of this effect of surface preparation it should not be assumed that poor line definition is a characteristic feature of reflection patterns, as has been stated by several authors. Patterns can be obtained with $K\alpha_1$ and $K\alpha_2$ clearly resolved and the black–white doublet contrast[32] clearly visible.

9.4.2 Excitation of Suitable Characteristic X-ray Wavelengths

The generation of a characteristic spectrum of X-rays from a particular region in the specimen is the central feature of the back reflection method applied to polycrystalline or multiphase materials. The maximum beam energy which can be used is limited because of the risk of having too high a background intensity from the X-ray continuum and there is a lower limit to the beam energy for a given specimen because the wavelengths generated must be short enough to obtain the diffraction effect. The d spacings which occur in most metallic crystals require that the wavelengths available must be shorter than approximately 2·75 Å, therefore the elements which will diffract their own K radiations start at α-titanium. When molybdenum is reached the K excitation level has reached 18 keV and to generate the K spectrum efficiently the beam voltage would have to be raised above 25 kV, which is not recommended. The required wavelength d-spacing relationship recurs again with the L spectra from the rare earths and good patterns are obtainable from tantalum and tungsten. This type of spectrum can be used until gold or lead, but at uranium the L excitation level has reached 17 keV and so becomes difficult to excite.

One can therefore divide the elements up into those which will easily produce X-ray wavelengths suitable for Kossel patterns and those which for a variety of reasons will not. For the second group it is necessary to deposit a layer on the surface of the specimen which will provide X-rays of a suitable wavelength to be diffracted by the crystals underneath. Also, to retain one of the main features of the microdiffraction method, it should still be possible to recognize where the pattern has come from. To achieve conditions as close as possible to the case where the source is in the crystal, the layer has to be thin enough for the source to be as close to the crystal as possible and yet thick enough to generate a 'point' source of X-rays. Fortunately all these conditions can be satisfied at the same time.

Since the most useful wavelengths for diffraction are between 1·5 and 2·0 Å then copper or manganese are suitable metals to evaporate on to the surface. It has been found experimentally that the optimum thickness of these layers corresponds to the 'depth of diffusion' which the electrons reach after penetrating into the target and this can be obtained from the results of Bishop[29] or from the formula of Cosslett.[33] For copper, this depth, the depth at which most ionization occurs, is 0·5 to 0·75 μm and with this thickness and conducted current imaging in the scanning system the underlying surface features can still be seen with beam energies of 25 keV. With 0·75 μm thickness of manganese on a specimen of iron the diffraction pattern was due solely to the MnKα radiation at a beam voltage of 20 kV but at 25 kV the pattern contained reflections of FeKα which showed that the source was extended from the evaporated layer into the underlying crystal. If the evaporated layer is too thick then the pattern is not complete in the centre, unless a high index plane is parallel to the surface, or nearly so. The reason for this effect is that the solid angle of the X-ray beam is reduced by being moved outside the diffracting crystal and, on the simple geometrical model, the planes do not extend far enough to produce a complete cone. An explanation of the effect in terms of the reciprocal lattice has been given by Mackay.[8] The loss of pattern increases with increasing thickness of the evaporated layer, so control of the thickness is important if the pattern is to be complete. If a surface layer has to be used to generate the necessary wavelength, it is probably better to use the transmitted pattern with thin specimens instead of thick specimens and the reflected pattern.

9.4.3 Instrumental Conditions

The patterns are recorded on standard X-ray film. The fine-grained types on a stiff polyester base produce the best results and have the lowest background colour, but satisfactory results with shorter exposure times can be obtained with the coarser-grained faster types of X-ray film. With the finest-grained film a current of 10^{-8} A in the specimen will produce a good pattern in between 5 and 7 min exposure from an iron crystal.

Currents of this value are higher than normal for scanning microscopy and other instrumental factors require adjustment. In particular the working distance needs to be greater than usual to achieve a sufficiently large solid angle at the film. There are two limitations on the working distance. The most restrictive is the design of the specimen stage which usually has only a limited Z movement. The second limitation is an electron-optical one. Increasing the working distance increases the spherical aberration, reduces the current which can be put into the source and increases its minimum size. However the angular aperture is also reduced so the increase in size is not as great as one might expect. In addition the image resolution requirements are not severe. The features examined are of the sizes observable with optical microscopes and patterns are diffuse or incomplete when generated from crystals $2 \mu m$ across. In the worst cases, with little interphase contrast in the image, the crystal might only be outlined by an etched boundary perhaps a tenth of this size. In these circumstances the resolution of the scanning image need be no better than 2000Å and would probably be adequate for most of the time at $0.5 \mu m$. Also the magnification is lowered by increasing the working distance so the resolution needed is lessened. The necessary conditions of sufficient current and adequate resolution at extended working distances seem to be easy to attain in several different scanning microscope instruments.

In the previous section there was mention of the need sometimes to deposit a surface layer on the specimen in order to generate X-rays of a suitable wavelength. In these circumstances the conducted electron mode of imaging will still produce an image of the underlying surface features and the higher currents needed to generate the pattern are helpful in this imaging mode while the relaxed resolution requirements suit this mode too.

9.4.4 Heating Effects of the Electron Beam

The heating effect of the electron beam has been noted several times in the literature on Kossel microdiffraction,[5,8,10] and for the measurement of plane spacings or lattice constants the magnitude of the effect must be known and reduced as much as possible. Yakowitz[10] described a semi-empirical approach originated by Morris for estimating the temperature rise and Hanneman et al.[5] used a formula for calculating the estimate. The main problem with these methods is to estimate the size of the region affected by the electron bombardment, which is contributing to the diffraction pattern and causing an apparent increase in the plane spacings. Hanneman et al. used a relatively large volume for their estimate and hence arrived at a very low temperature rise. Mackay[8] showed that their result (for pure nickel) was probably in error because if the X-rays were generated in a target separated from the specimen, a result close to the result obtained by Debye–Scherrer methods was achieved. The result obtained by Hanneman et al. was an expansion equivalent to a temperature rise of $70 \,°C$

and the result obtained by Heise[11] is consistent with a temperature rise of 130 °C.

An alternative method for thick specimens has been used by Harris[34] who measured the plane spacings or the lattice parameters at three or more values of the electron beam power. Results show that over the range of power used in practice, the expansion is proportional to the power. Extrapolation to zero power therefore should give the value of the lattice parameters which would be obtained if the heating effect were absent. Using the standard expression for the linear expansion of a solid, the largest change in plane spacing measured by Harris was equivalent to a temperature rise of 86 °C.

9.5   INTERPRETATION OF THE PATTERN

The structure of the Kossel pattern is the same as a Laue back reflection pattern; it is a set of conic sections in gnomonic projection. The major differences are that the axes of the cones are plane normals and not zone axes, the intersections between the cones are not rational indices and the curvatures of the conic sections are a function of the plane spacing and the X-ray wavelength. Some methods of interpretation therefore are analogous to those used for Laue patterns—indexing and orientation measurements, for example—while others can make use of the extra information provided by the plane spacing effects.

9.5.1 Indexing
The amount of information in a Kossel pattern is large, and in a small region the pattern can look complicated, but indexing the curves is straightforward and the results are reliable because any deductions can be cross-checked in several ways. One can therefore reduce the magnitude of the 'indexing problem' which causes so much trouble with Debye–Scherrer patterns from low symmetry crystals. In particular, overlaps are impossible. Kossel curves can be indexed by one or other, or a combination of, the pattern symmetry, the line curvature, the intersections of curves and their relative positions. Results from research carried out into this problem in the author's laboratory and under his direction have been described by Harris.[35]

Directions with low-order rational indices that are axes of symmetry are the points of intersection of so-called mirror lines in the pattern.[14] Mirror lines can be drawn so that part of the pattern on one side is a reflection of the

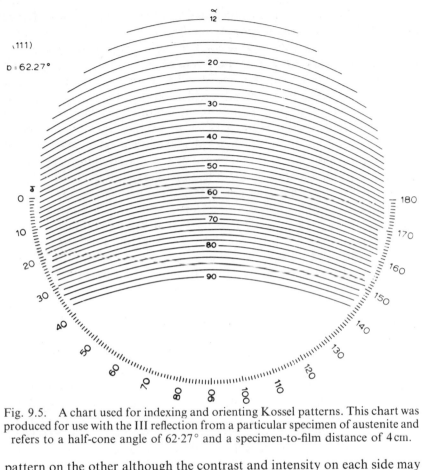

Fig. 9.5.   A chart used for indexing and orienting Kossel patterns. This chart was produced for use with the III reflection from a particular specimen of austenite and refers to a half-cone angle of 62·27° and a specimen-to-film distance of 4 cm.

pattern on the other although the contrast and intensity on each side may not be the same. The establishment of the pattern symmetry in this way is therefore a useful first step. If the solid angle of the pattern is large enough (approaching 90°) the pattern can be seen as an approximation to a stereographic projection, and working from symmetry and relative curvature estimated visually, most of the pattern can be indexed by inspection if the crystal symmetry is high enough to cut down the number of reflections.

One can also use the curvature of the lines more directly and combine this with a measurement of the angular position of the cone axes by using the Kossel analogues to Greninger charts.[36,37] An example is shown in Fig. 9.5. The charts have pre-calculated conic sections representing different degrees of tilt of a cone of fixed apex angle ($= 2(90 - \theta)$). The procedure is essentially a curve-fitting procedure and so the cone angles should be for low index planes (i.e. around $30°\theta$). It is sufficient if a set of charts is available with increments of $2°$ in θ because interpolation is easy if the curvature is not too great and so, for a fixed distance between specimen and film, the number of charts required would be no more than about ten. If the lattice constants are known, therefore, the use of these charts indexes the reflection from its curvature and fixes the angular co-ordinates of its plane normal. Several cones can be indexed and plotted on a stereographic projection in a few minutes. The interplanar angles derived from these plots are accurate[24] despite the apparent low precision.

If the crystal is not known, more information needs to be accumulated and the intersections between the cones then become valuable. In particular, multiple intersections of type II depend only on the lattice geometry and have the additional property that the total number is a characteristic of the crystal type.[35] Further, one can subdivide the multiple intersections into classes of five-, four- or three-fold intersections (i.e. the intersections of five, four or three Kossel lines). The number in each class is then characteristic of the crystal and the planes forming the intersection. Tables can then be prepared in advance showing the differences one would expect between different structures.

9.5.2 The Determination of Lattice Constants or Plane Spacings

The main application of the Kossel diffraction technique from its discovery in 1935 until recently has been the derivation of lattice constants or plane spacings from measurements on the recorded pattern. Early investigators derived values of the lattice parameters of crystals by measuring the relative positions of two or more Kossel curves and these can be called 'lattice constant methods'. The last ten years have seen the growth of a different approach to the problem in which the pattern is studied one curve at a time and the spacing of the crystal planes responsible for each curve is then derived. We shall call these 'plane spacing methods'. Lattice constant methods rely on the geometry of the crystal and the growth of plane spacing methods is due to the increase in the study of situations such as strain measurement where the crystal geometry is disturbed and the former approach is no longer valid. The papers by Yakowitz[10] and Tixier and Waché[21] are extensive reviews of methods for obtaining lattice constants or plane spacings. The paper by Yakowitz is valuable as a critical assessment of the earlier methods and the other paper is a valuable survey of the more modern methods. Because of the existence of these reviews we can be more

TABLE 9.2
Summarized Methods for Obtaining Crystal Constants

Method	Initial information	Max. precision claimed
(a) *Lattice constant*		
Kossel[3]	1 or 5, 3, 8, 9	1 in 7 000
Lens	1, 2, 3, 5, 8, 9	See Yakowitz[10]
Accidental intersections[38]	3, 6	1 in 10^6
Tixier and Waché[21]	3, 4, 8	1 in 10^4
(b) *Plane spacings*		
Ellis et al.[50]	5, 7, 8	2 in 10^4
Fisher and Harris[51]	7, 8	5 in 10^4
Morris[17]	1, 2, 8	
Harris and Kirkham[19]	8	1 in 10^3

Information requirements: 1. Distance between source and film. 2. Pattern centre. 3. Crystal geometry or indexed pattern. 4. Approximate lattice parameters. 5. Special orientation of crystals. 6. Accidental intersections of three or more curves at a point. 7. Several exposures of the pattern. 8. Precise measurements on the film. 9. Two curves overlapping or nearly overlapping.

Precise values of the wavelength are always required particularly for the accidental intersection methods.

selective but the major methods and their experimental requirements are summarized in Table 9.2. It will be noted that there is a wide choice of methods and a selection of a method can be made depending on the situation and how much information is available. This wide choice of methods for the interpretation of a pattern, by comparison with more conventional X-ray techniques, applies also to orientation determinations as we shall see later. The variety of methods arises because of the very large amount of information in the pattern and is a characteristic feature of Kossel pattern techniques. Investigators used to conventional techniques, with generally one method of interpretation for one type of pattern, have therefore to become used to the situation where there is a multiplicity of methods for the interpretation of what is really a multi-purpose pattern.

9.5.3 Lattice Constants
The quickest method of determining the lattice constants is probably by the use of accidental intersections. We have seen in a previous section that multiple intersections of Kossel curves can be either of geometrical origin,

or 'accidental' due to a combination of the effects of the X-ray wavelength and the crystal geometry (Type III, Fig. 9.3). Accidental intersections are distinguished from the others by not being formed in pairs and hence the three curves meet at one point and at no other. The unique nature of such an event results in a powerful method of establishing the lattice constants.

Consider three curves in a cubic diffraction pattern coming from crystal planes with Miller indices $(h_ik_il_i)$ ($i = 1, 2$ or 3). The *Kossel* plane associated with each diffraction curve is given by

$$\frac{h_iX}{a} + \frac{k_iY}{a} + \frac{l_iZ}{a} = \frac{\lambda}{2}\frac{1}{d^2}$$

where a is the lattice constant, λ is the wavelength and d is the plane spacing expressible in terms of $h_ik_il_i$ and a. $X^2 + Y^2 + Z^2 = 1$ if the Kossel planes are to meet at a point. There are therefore four equations and four unknowns: X, Y, Z and a, which can be solved to find a. The lattice parameter can therefore be calculated from a single accidental intersection as soon as the three curves are indexed.

To extend this method to crystals with lower symmetries it is necessary to use more than one accidental intersection. Schwarzenberger[38] used two intersections and eight equations for hexagonal crystals and the same would apply to tetragonal cases. Three accidental intersections are needed for orthorhombic crystals and this method will be used as an example to determine the lattice constants of cementite. In these more difficult cases it is better to assume approximate values for the constants and find the correct values by iteration using a Taylor series.

The cementite was part of the eutectic of a white cast iron of composition $3\cdot6\%\,C$, $1\%\,Si$, $12\%\,Mn$. Three accidental intersections were observed: A, formed by $103\,FeK\alpha_1$, $1\bar{1}3\,FeK\alpha_1$, $\bar{2}33\,FeK\alpha_1$; B, formed by $0\bar{2}3\,FeK\alpha_1$, $\bar{1}13\,FeK\alpha_1$, $3\bar{1}3\,FeK\alpha_1$; and C, formed by $0\bar{2}3\,FeK\alpha_1$, $113\,FeK\alpha_1$, $103\,MnK\alpha_1$. The values of the constants a, b and c calculated from this information were $a = 4\cdot5455\,Å$, $b = 5\cdot0785\,Å$, $c = 6\cdot6990\,Å$. These compare with the values for pure cementite obtained by Fasiska and Jeffrey:[39] $a = 4\cdot5248\,Å$, $b = 5\cdot0896\,Å$, $c = 6\cdot7443\,Å$. The values from the manganese cementite have not been corrected for the heating effect, which would make them too high by approximately $0\cdot001\,Å$.

It should be noted that the occurrence of accidental intersections is fortuitous and they may not be observed in some patterns, although deliberately increasing the number of wavelengths in the pattern will improve the chances. Patterns from tantalum, austenite, and cementite show accidental intersections and the method can be extended to the case where the three curves in the cubic case just miss forming an accidental intersection.[7]

9.5.4 Crystal Plane Spacings

All plane spacing methods depend on the calculation of the semi-apex angle (α) of the diffraction cone where α is the complement of the Bragg angle. The most general method relies on establishing the equation of the conic section on the film by obtaining the co-ordinates of several points along its length and then fitting the best curve to this data. This approach was originally used by Morris[17] and further improved by Yakowitz[10] but they required that the pattern centre and the distance from the source to the film be known. Harris and Kirkham[19] developed a suggestion by Bevis[18] to overcome these requirements and their method requires only the co-ordinates of points along the Kossel curve. The more points used the better, and up to twenty are usually used. It is not necessary to know the indices to determine only the d-spacing, but it is necessary if the lattice constants are to be determined. Only two curves are needed for plane spacing measurements but, since the method can also be used to establish the orientation when three curves are used, it is better to use three curves every time. Film shrinkage effects can be ignored because, if they are uniform, they appear as an apparent change in the specimen-to-film distance.

In its original form the method of Harris and Kirkham was of relatively low precision. Biggin and Dingley[20] have developed a method of superimposing artificial conic sections on the pattern, which enables the pattern centre to be determined with greater reliability. This technique improves the Harris and Kirkham method and it then becomes probably the most generally applicable approach.

9.6   THE ORIENTATION OF MICRO-CRYSTALS

The orientation or relative orientation of crystals in metallographic specimens has been a most fruitful field of application for the back reflection Kossel method. A knowledge of crystal and habit plane relationships in studies of phase transformations is essential to a complete description of these processes, and Kossel microdiffraction by either reflection or transmission is a valuable technique for the following reasons: suitable sites are easy to locate, the results are unambiguous and patterns can be obtained from a large enough number of sites to produce a statistically significant result. These advantages have been exploited in several successful major studies, using the reflection method in the Department of Metallurgy and Materials Science in the University of Liverpool. The main applications have been in the following fields: graphite precipitation in white cast irons,[40,41] martensitic transformations in Fe–Ni and Fe–Ni–C alloys,[42,43] recrystallization in pure iron,[44] and growth of pro-eutectoid ferrite.[46,47] At the University of Sussex, Bellier[45] used the transmission method to study recrystallization in pure aluminium.

There is an added advantage in using the reflection method and thick specimens in that two surface trace analyses can be carried out to obtain precipitate habit planes or grain boundary surfaces in terms of the orientations of the crystals forming the boundary. This can be achieved with crystals of a few microns in size. Hence a large number of possible applications exist with opportunities to obtain a complete crystallographic description of metallographic structures in crystalline materials.

The orientation of a crystal may need several parameters to define it, and the amount of information needed to be useful can vary considerably. One can divide the information requirements into two main sections initially: one where the diffraction pattern has to be related to surface traces and the other where no surface features are included in the result. The simplest case in the second section is when only the Miller indices of the surface normal are required. Then no extra information is required other than that on one film, for a single orientation, or on two films for relative orientations. If the crystallographic directions of X and Y axes in the plane of the specimen surface are needed in addition to the Z axis, then for relative orientations some standard reference system is required to make comparisons between patterns from the two crystals. The simplest form this could take would be north south and east–west axes on the recording film.

If the diffraction results are to be related to surface traces then there is an essential requirement for a reference system between the specimen surface and the recording film in addition to the film reference system. Finally there will be the results from angular measurements of the surface trace, with respect to the surface-to-film reference, to combine with the XYZ directions in the pattern, and the trace angles could be from measurements on either one or two surfaces in the specimen.

The methods of pattern interpretation can also be divided into two main classes: firstly, there are those which use a stereographic projection to establish the final result, and secondly there are those which need only measurements made on the recording film.

The simplest approach to orienting the pattern is to convert the gnomonic projection on the film into a stereographic projection and treat the result like the similar result from a Laue X-ray diffraction pattern. Peters and Ogilvie[12] were the first to apply this approach and their method can always be used. While it is not recommended for routine use it is a valuable standby. It works best with complete cones, which will be from high index planes, and the measurements are made easier with the use of simple jigs for holding the film.[28] The results are not very accurate, judged by the values of interplanar angles determined in this way and the Bragg angles of the cones have a possible error of $\pm 1°$.

This approach is improved considerably by the use of the pre-calculated charts referred to earlier and their application has been extended by the realization that the charts can also be used to establish the pattern centre.[12] This enables the chart method to be applied to different specimen–film arrangements than was originally intended. The chart method is therefore the recommended one for rapid orientation on a routine basis if a computer is not available.

Relative orientations can be determined by constructing the symmetry axes on the pattern from intersections of the lines of symmetry in the pattern and then simply superimposing the patterns making use of some simple reference marks produced by the camera. This method was used by Harris et al.[48] for the orientation of tantalum oxide platelets growing into tantalum.

Analytical solutions to determining orientations achieve the orientation by calculation with a computer. The general interpretation method of Harris and Kirkham[19] referred to earlier can be used for orientation studies but requires many precise measurements along each of three curves. Fewer measurements are needed in the method derived by Ryder et al.[14–16] and this method has the same important advantage of not needing to know the pattern centre.

In the method of Ryder et al. the *film* normal is calculated as the result of the vector product of vectors with known components in the plane of the film. These vectors are chosen by joining, on the film, points with known indices. Four such points are chosen and they can either be intersections whose indices can be calculated, or intersections of lines of symmetry in the pattern. A fifth point is required and this is the intersection of the two lines joining the pairs of opposite points. The original references should be consulted for the simple equations used. It may not be necessary to enlarge the pattern by optical projection to make the measurements, as the originators did. If the points used are intersections of Kossel curves then their co-ordinate positions could be established with a two-dimensional travelling microscope and the lines joining them in the plane of the film and their intersection could be established by analytical geometry. If four other sets of co-ordinates could describe the film reference axis in the same co-ordinate system, then this would be all the information one would need for an orientation determination. The result could also be related to a single surface trace angle through a film and specimen reference system. This method is probably the best of the analytical methods available and has been extended by Tixier and Waché[21] to be used as the basis for a method of determining lattice constants to a high precision.

9.6.1 Two-surface Trace Determination

The use of two-surface trace analysis is a standard procedure when studies of large single crystals are made. The same kind of information is needed from the crystal sizes accessible to Kossel microdiffraction and the problem is to achieve this without losing the capability to examine many sites. Two techniques have been successful in this respect.

The first, which is usually called serial sectioning, determines the angle a boundary makes with a surface by measuring the shift in position as the surface is polished away by known amounts. The area of interest is recorded as a transparent positive at each polishing stage and the biggest problem is to superimpose the images with sufficient precision to make the shift measurement precise. One way of achieving this and measuring the amount removed by polishing is to use micro-hardness indentations in the area recorded. The amount removed can be calculated from the change in dimensions of the impressions, provided the sides are plane, and the images of the indentation can be superimposed to achieve the correct registration.

The precision of the measurement of the depth removed by this means is approximately 6 % and the possible error in the angle of the interface is 6° as a maximum and can be less under favourable conditions, as was shown by the results of Bevis and Swindells.[13] The other method used by Rowlands *et al.*[43] is useful when the features being examined are relatively long. The surface of the specimen is ground and polished to a roof shape with an apex angle of ∼ 170°. The angular relationship between the two sides can easily be determined by a simple tilting stage on an optical microscope. The ridge of the roof can be made to run through several features at once (e.g. martensite plates) and then while one surface is used for the diffraction pattern, photographs of both surfaces are used to obtain the trace angles. The correct register between the photographs can easily be achieved if a linear contamination mark deposited by the electron probe is made by scanning the probe across the ridge between the two surfaces.

9.6.2 Surface Reference and Image Inversions

One can see from previous sections that to achieve the maximum amount of information from the surface, some reference between the surface and the pattern is needed in order to relate measurements made in both. The reference needs to be on the same scale as the features being examined and should not damage the surface in any way. This problem was solved by Bevis and Swindells.[13]

The simplest reference line on the surface is a line contamination mark made by scanning the beam along one axis of the image frame. The alignment of this line with an external reference frame is achieved by using the stage shifts to align the image direction with the stage direction. It is helpful if the scanning image frame can be rotated to make the two directions coincide. The film reference marks can be non-identical indentations in the circumference edges of the film cassette and then the cassette can be located mechanically in a fixed position with respect to the stage axes. Then the image axes and the film axes are related via the stage axes.

Because of the distortions and inversions present in most scanning images, surface trace angles are measured more accurately on an optical metallograph. The linear contamination mark then serves as the reference mark, but to allow for inversions it should be made unsymmetrical with a spot contamination mark to one side of one end of the line. The film must be marked so that it is always viewed from one side and this must be chosen with respect to the specimen surface in a consistent manner. Image inversions in the scanning image can be established with the use of the stage shifts and the same can be done with the metallograph image. The existence of inversions is not always realized and even when established, they must be checked periodically. In two apparently identical instruments used for Kossel microdiffraction the scanning image of one is the mirror image of the other, due to differences in the electrical connections.

9.7 CONCLUSIONS

Kossel X-ray diffraction from thick, crystalline materials should not be regarded as a specialized technique with limited applications and possible only in specially constructed apparatus. Kossel patterns should be obtainable from any crystalline substance whose crystal size is larger than $5\,\mu m$ and the patterns can be interpreted in several, straightforward ways.

The focussed electron beam used to generate a diffraction pattern from crystals of this size, or larger, is conveniently provided by an electron probe microanalyser or scanning electron microscope.

The possibility that Kossel X-ray diffraction could be generally available in this way leads to the conclusion that the use of conventional methods should be reconsidered. For example, there would appear to be little point in reducing a polycrystalline substance to powder in order to establish its lattice constants by the Debye–Scherrer method if the same information could be obtained with high precision from the polycrystalline solid. Further, if the polycrystalline solid contained more than one phase, the Kossel microdiffraction technique would be a more logical method to use, because each phase could be studied separately instead of mixed together.

9.8 ACKNOWLEDGEMENTS

The development of the back reflection Kossel microdiffraction technique at the University of Liverpool has been generously supported by the Science Research Council by research grants to the author. Throughout the whole development period considerable contributions have been made by Dr M. Bevis and his students, and it is a pleasure to acknowledge the many helpful and important discussions which have taken place during a fruitful collaboration.

REFERENCES

1. R. Castaing, Ph.D. Thesis, University of Paris, 1952 (ONERA Publication No. 55).
2. W. Kossel, V. Loeck and H. Voges, *Zeit. für Physik*, 1935, **94,** 139.
3. W. Kossel and H. Voges, *Ann. der Physik*, 1935, **23,** 677.
4. P. Gielen, H. Yakowitz, D. Ganow and R. E. Ogilvie, *J. Appl. Phys.*, 1965, **36,** 773.
5. R. E. Hanneman, R. E. Ogilvie and A. Mordrzewski, *J. Appl. Phys.*, 1962, **33,** 1429.
6. W. Kossel, *Ann. der Physik*, 1936, **25,** 512.
7. K. Lonsdale, *Phil. Trans. Roy. Soc. A*, 1947, **240,** 219.
8. K. J. H. Mackay, in *Optique des Rayons X et Microanalyse* (eds. R. Castaing, P. Deschamps and J. Philibert), p. 544. Hermann, Paris, 1966.
9. J. Z. Frazer and G. Arrhenius, in *Optique des Rayons X et Microanalyse* (eds. R. Castaing, P. Deschamps and J. Philibert), p. 516. Hermann, Paris, 1966.

10. H. Yakowitz, in *Advances in Electronics and Electron Physics*, Supplement 6, p. 361. Academic Press, New York, 1969.
11. B. H. Heise, *J. Appl. Phys.*, 1962, **33**, 938.
12. E. T. Peters and R. E. Ogilvie, *Trans. AIME*, 1965, **233**, 89.
13. M. Bevis and N. Swindells, *phys. stat. sol.*, 1967, **20**, 197.
14. P. L. Ryder, H. Hälbig and W. Pitsch, *Mikrochim. Acta*, 1967, Supplement II, 123.
15. P. L. Ryder, H. Hälbig and W. Pitsch, *Mikrochim. Acta*, 1968, Supplement II, 201.
16. P. L. Ryder, H. Hälbig and W. Pitsch, in *X-ray Optics and Microanalysis* (eds. G. Möllenstedt and K. H. Gaukler), p. 388. Springer, Berlin, 1969.
17. W. G. Morris, *J. Appl. Phys.*, 1968, **39**, 1813.
18. M. J. Bevis, E. O. Fearon and P. Rowlands, *phys. stat. sol.* (a), 1970, **1**, 653.
19. N. Harris and A. J. Kirkham, *J. Appl. Cryst.*, 1971, **4**, 232.
20. S. Biggin and D. J. Dingley, *J. Appl. Cryst.*, 1977, **10**, 376.
21. R. Tixier and C. Waché, *J. Appl. Cryst.*, 1970, **3**, 466.
22. G. G. Hall, *Matrices and Tensors*. Pergamon, Oxford, 1963.
23. *International Tables for X-ray Crystallography*, Vol. II, p. 60. Kynoch Press, Birmingham, 1959.
24. N. Swindells, in *X-ray Optics and Microanalysis* (eds. G. Möllenstedt and K. H. Gaukler), p. 383. Springer, Berlin, 1969.
25. F. Maurice, J. Philibert, R. Seguin and R. Tixier, *J. Appl. Cryst.*, 1975, **8**, 287.
26. M. Bevis and N. Swindells, *J. Mat. Sci.*, 1973, **8**, 898.
27. N. Swindells and J. C. Ruckman, *Scanning Electron Microscopy*, Conference Series 18, p. 302. Institute of Physics, London, 1973.
28. D. L. Veith and H. Yakowitz, *J. Res. Nat. Bur. Stds*, 1967, **71C**, 313.
29. H. E. Bishop, *Proc. Phys. Soc.*, 1965, **85**, 855.
30. H. Kanter, *Ann. der Physik*, 1957, **20**, 144.
31. L. E. Samuels, *Metallographic Polishing by Mechanical Methods*. Pitman, London, 1971.
32. M. T. F. von Laue, *Röntgenstrahlinterferenzen*, p. 367. Akademische Verlagsgesellschaft Geest and Portig K-G Leipzig, Zwt. Aufl., 1948.
33. V. E. Cosslett, *Brit. J. Appl. Phys.*, 1964, **15**, 107.
34. N. Harris, *J. Mat. Sci.*, 1974, **9**, 1211.
35. N. Harris, *J. Mat. Sci.*, 1975, **10**, 279.
36. P. C. Rowlands and M. Bevis, *phys. stat. sol.*, 1968, **26**, K25.
37. G. Ferran and R. A. Wood, *J. Appl. Cryst.*, 1970, **3**, 419.
38. D. R. Schwarzenberger, *Phil. Mag.*, 1959, **4**, 1242.
39. E. J. Fasiska and G. A. Jeffrey, *Acta Cryst.*, 1965, **19**, 463.
40. N. Swindells and J. Burke, in *The Mechanisms of Phase Transformation in Metals*, p. 92. Institute of Metals, London, 1969.
41. N. Swindells, J. D. Avery and N. Harris, in *The Metallurgy of Cast Iron* (eds. B. Lux, I. Minkoff and F. Mollard). Georgi, St Saphorin, 1975.
42. P. C. Rowlands, E. O. Fearon and M. Bevis, *J. Mat. Sci.*, 1970, **5**, 769.
43. P. C. Rowlands, E. O. Fearon and M. Bevis, *Trans. AIME*, 1968, **242**, 1559.
44. A. Dunn, Ph.D. Thesis, University of Liverpool, 1971.
45. S. P. Bellier, Ph.D. Thesis, University of Sussex, 1971.
46. A. D. King and T. Bell, *Metal Science*, 1974, **8**, 253.

47. A. D. King and T. Bell, *Metal Science*, 1975, **9**, 1419.
48. N. Harris, A. Taylor and J. Stringer, *Acta Met.*, 1973, **21**, 1677.
49. D. J. Dingley and S. Biggin, *Scanning Electron Microscopy, Conference Series* 18, p. 314. Institute of Physics, London, 1973.
50. T. Ellis, L. F. Nanni, A. Shrier, S. Weissmann, G. E. Padawar and N. Hosokawa, *J. Appl. Phys.*, 1964, **35**, 3364.
51. D. Fisher and N. Harris, in *X-ray Optics and Microanalysis* (eds. G. Möllenstedt and K. H. Gaukler), p. 369. Springer, Berlin, 1969.

Index